EINSTEIN'S 1912 MANUSCRIPT ON THE SPECIAL THEORY OF RELATIVITY

$$E = \frac{mc^2}{\sqrt{1 - \frac{q^2}{c^2}}}$$

To Our Mariner Sands Einstein

Happy Birthday

From Your

Tennis Pro Chris

EINSTEIN'S 1912 MANUSCRIPT ON THE SPECIAL THEORY OF RELATIVITY

George Braziller, Publishers
in association with the
Edmond J. Safra Philanthropic Foundation

Einstein's original 1912 Manuscript on the Special Theory of Relativity was acquired by Edmond J. Safra through his foundation and generously donated to the Israel Museum in Jerusalem, where it continues to reside.

First published by George Braziller, Inc., in association with the Jacob E. Safra Philanthropic Foundation and the Israel Museum, Jerusalem, in 1996.

For information, please address the publisher:
George Braziller, Inc., 171 Madison Avenue, New York, New York 10016

Library of Congress Cataloguing-in-Publication data:

Einstein, Albert, 1879–1955.
[Spezielle Relativitätstheorie. English & German]
Einstein's 1912 manuscript on the special theory of relativity : a facsimile.
p. cm.
English and German text appear on facing pages
Includes bibliographical references.
ISBN 0-8076-1417-3 (cloth)
ISBN 0-8076-1532-3 (paper)
1. Special relativity (Physics) 2. Einstein, Albert, 1879–1955—Manuscripts—Facsimiles. 3. Manuscripts, German—Facsimiles. 4. Einstein, Albert, 1879–1955—Biography. 5. Physicists—Biography. I. Title.
QC173.65.E3613 1996 96-27567
530.1 ' 1—dc20 CIP

Designed by Philip Grushkin

Printed and bound in Hong Kong through Asia Pacific Offset, Inc.

Revised Edition, 2003

CONTENTS

ALBERT EINSTEIN
Scientist, Humanist, Zionist

On November 16, 1952, Israeli Prime Minister David Ben-Gurion sent an urgent cable to Israel's ambassador in Washington, Mr. Abba Eban, instructing him to immediately inquire whether Albert Einstein would agree to be the president of the State of Israel. Writing about this episode years after, Itzhak Navon, then Ben-Gurion's secretary and later an Israeli president, recalled that after sending this cable Ben-Gurion asked, "What do we do if he says yes? I had to offer it to him, but if he accepts we are in trouble." Einstein did not disappoint Ben-Gurion. He rejected the offer, saying, "I know little about nature, but I know almost nothing about human beings."

Why did Ben-Gurion think he had to offer this position to Einstein, and why was he so concerned that he might accept? Ben-Gurion considered Einstein to be the greatest Jew on earth, who not only did not deny his Jewishness but was proud of it. By offering the presidency to Einstein, Ben-Gurion wanted to demonstrate that culture and science are the pinnacle of achievement for the Jewish people. On the other hand, Ben-Gurion was probably worried about Einstein's uncompromising opinions on controversial political issues.

Einstein was first and foremost the architect and engineer of a new understanding of the physical world—the most revolutionary innovator since Newton. Although his fame stems from his achievements in science, great respect and attention were also paid to his person and to his opinions outside the world of physics.

Einstein was a prolific writer. In correspondence with peers, in numerous articles, and in public addresses, he expressed forceful and unswerving opinions on a variety of public, political, and moral issues such as nationality and nationalism, war and peace, human liberty and dignity, and launched an untiring attack on any form of discrimination. Bluntly expressed, controversial, often considered simpleminded and naive, his positions nevertheless had a significant impact.

The influence of Einstein's contributions on so many branches of physics is such that if one wanted to describe its full extent, it would be hard to know where to begin. His work and discoveries in statistical physics, in the modern physics of quanta and radiation, and in the understanding of the photoelectric effect are so fundamental that each achievement alone would have guaranteed him a prominent place in the history of physics. But what brought him unprecedented fame outside his own discipline is undoubtedly his theory of relativity, which revolutionized the old, established Newtonian picture of space, time, and gravitation.

Einstein regarded nature as a rational system that could be analyzed scientifically to establish objective knowledge. He compared nature to a vast mystery story and scientists to readers who, for each episode, seek solutions that are consistent with the available facts. They are like Sherlock Holmes, who, on the basis of apparently unrelated clues, find an ingenious solution through the application of pure thought. In the same way, Einstein pieced together certain clues and contradictions encountered in physics toward the end of the nineteenth century, and in 1905, he laid down two principles:

1. The velocity of light does not depend on the velocity of the source or of the measuring device.

2. To explain the second principle, a metaphor used by Einstein himself helps. If we suppose that we are sitting in a uniformly moving train with no windows, then there is no phenomenon and no experiment that will allow us to detect the motion of the train.

These two deceptively simple statements have profound consequences that affect all of the physical sciences. They are the basis of the Special Theory of Relativity. By pure logic and mathematical derivation, one conclusion follows another—thus two observers moving at different velocities and measuring the length of the same stick will obtain different answers; looking at the same clock, they will measure different times; and looking at two events, they will sometimes disagree on which precedes the other. Further analysis leads to the equivalence of mass and energy, as reflected by the most renowned combination of five symbols: $E=mc^2$. It is science's most famous equation, it appears on stamps and in commercials—it is the most famous, but also the least understood. A long but straight road leads from this formula to Hiroshima. Development of sophisticated technology was required, but the principle on which the atomic bomb is based is expressed by this combination of five symbols.

All this constitutes Einstein's Special Theory of Relativity, which is summarized in his unpublished 1912 manuscript now displayed in the Israel Museum thanks to the initiative of Lily and Edmond Safra and to the generosity of the Jacob E. Safra Philanthropic Foundation.

The Special Theory was followed by the General Theory of Relativity in which the metaphor of the uniformly moving train is replaced with the metaphor of a falling elevator. Again, mathematical analysis leads from simple basic principles to profound consequences concerning the structure of the universe and to a new theory of gravitation.

In March 1979 Jerusalem's Mayor Teddy Kollek and Professor Aryeh Dvoretsky, the president of the Israel Academy of Science and Humanities, convened the Jerusalem Einstein Centennial Symposium to mark the 100th anniversary of Albert Einstein's birth. The symposium brought to Jerusalem a distinguished group of scientists and scholars who were able to shed light on many aspects of Einstein's achievements and personality. On that occasion, Sir Isaiah Berlin spoke about Einstein and Israel. He stated that there can be no doubt about the relevance of Einstein's views to one of the most positive political phenomena of our time: " Einstein lent the *prestige mondial* of his great name, and in fact gave his heart, to the movement which created the state of Israel." Even though he had frequent arguments with Zionist leaders, particularly with regard to attitudes toward the Arabs, Einstein's support of the Zionist movement and of the State of Israel was lifelong. Toward the Arabs, he always demanded human decency, especially from his own people. Yet he believed in the basic principles of Zionism and supported the State of Israel until the end of his life.

This conviction of his was a reflection neither of his childhood and family background, nor of his general principles, which deplored any form of nationalism and sectarianism, and certainly not of the prevailing opinions of his intellectual milieu. In fact, many considered his attitude toward Zionism as one of the most contradictory aspects of his personality.

However, this attitude must be examined in the context of Einstein's understanding of the Jewish religion and civilization, which he interpreted as an eternal quest for universal human values, social justice, and the pursuit of truth through a long tradition of learning. His understanding of the Jewish legacy was that the sovereignty of an individual plays a central role, that the function of society is to enable individuals to develop their identity and their intellectual and moral potential, and that the framework

of statehood exists only in order to serve society. Judaism and Zionism were, in his mind, frameworks in which universal values are embodied. It was within such a context that he was convinced that circumstances made the creation of the State of Israel inevitable; he believed that a state would be the proper arena in which those worthy attributes of Jewish existence and creativity could be revitalized and flourish.

In 1933, Einstein wrote, "Palestine is not primarily a place of refuge for the Jews of eastern Europe, but the embodiment of the re-awakening of the corporate spirit of the entire Jewish nation." Since Einstein gave public support to the Zionist enterprise, he felt himself a partner to all aspects of its implementation, and he felt a moral responsibility on topics beyond his own involvement. His partnership in the Zionist program was most evident in his association with the Hebrew University of Jerusalem. Already in 1921 he had clearly voiced his opinion on the desired structure and mission of this university. He became an active member of its Board of Governors. He perceived the Hebrew University as the stage on which the inventiveness of the Jewish people and the Jewish quest for learning would come to prominent fulfillment. Einstein had fierce arguments with the administration of the Hebrew University in the late 1920s and 1930s, and it was these disagreements that he cited as the reason for not joining the university at that time. However, they did not affect his interest in the university—an interest that continued until the end of his life.

In his last will and testament, Einstein bequeathed his literary estate and personal papers to the Hebrew University of Jerusalem. The Albert Einstein Archives at the Jewish and National University Library constitute a cultural asset of supreme importance for humankind and a national asset of principal significance for the Jewish people. Its holdings are simply unique: they consist of the numerous manuscripts and prolific correspondence preserved by Einstein himself during his career as well as a large amount of additional material collected after his death. His personal papers represent the most important accumulation of material by and about Einstein in the world. They reflect the multifaceted aspects of his scientific work, political activities, and private life.

It is a puzzle in itself that so many idolized Einstein for an achievement of which they understood almost nothing. On the other hand, it may be a tribute to humankind that an individual can be so honored for what is generally only sensed as an outstanding intellectual breakthrough. Likewise, it is a tribute to Lily and Edmond Safra that Einstein's name still generates such thrill, excitement, and interest. Through the purchase of the 1912 Einstein manuscript on special relativity, not only has the Jacob E. Safra Philanthropic Foundation given this testimony of Einstein's genius an eternal home in Jerusalem, where it rightfully belongs, but the Foundation has also provided an opportunity to shed again some light on Einstein's many-faceted personality.

Professor Hanoch Gutfreund
Andre Aisenstadt Professor of Theoretical Physics
President of the Hebrew University of Jerusalem

In Einstein's Own Words

"The pursuit of knowledge for its own sake, an almost fanatical love of justice, and the desire for personal independence—these are the features of the Jewish tradition which make me thank my stars that I belong to it."

"I want to know how God created this world. I am not interested in this or that phenomenon, in the spectrum of this or that element; I want to know His thoughts; the rest are details."

"For the most part I do the thing which my own nature drives me to do. It is embarrassing to earn so much respect and love for it."

"Science and art are the only effective messengers for peace. They tear down national barriers, they are far better assurances of international understanding than treaties."

"Man tries to make for himself in the fashion that suits him best a simplified and intelligent picture of the world; he then tries to some extent to substitute this cosmos of his for the world of experiences, and thus to overcome it. This is what the painter, the poet, the speculative philosopher, and the natural scientist do, each in his own fashion."

"True scientific thought is not possible without faith in the inner harmony of our universe, and from this axiom I developed my theory of relativity."

"Falling in love is not at all the most stupid thing that people can do—but gravitation cannot be held responsible for it."

"Put your hand on a hot stove for a minute, and it seems like an hour. Sit with a pretty girl for an hour, and it seems like a minute. That's relativity."

"Great spirits have always encountered violent opposition from mediocre minds."

"You must be able to distinguish between what is true and what is real."

"Imagination is more important than knowledge. Knowledge is limited. Imagination encircles the world."

"The most beautiful experience we can have is the mysterious. . . . I am satisfied with the mystery of the eternity of life."

"Common sense is the collection of prejudices acquired by age 18."

"Only two things are infinite, the universe and human stupidity, and I'm not sure about the former."

"Nationalism is an infantile disease. It is the measles of mankind."

"Violence sometimes may have cleared away obstructions quickly, but it never has proven itself creative."

"Warfare cannot be humanized. It can only be abolished."

"I do not know with what weapons World War III will be fought, but World War IV will be fought with sticks and stones."

Chronology of the Life of Albert Einstein

1879 Born on March 14 in Ulm, Germany.

1903 Marries Mileva Maric, the mother of his one-year-old daughter.

1905 Receives his Ph.D. in physics from the University of Zurich. Unable to obtain an academic position, he starts working in the Swiss Patent Office in Bern as preliminary examiner of patent applications. Left with ample free time, he publishes four revolutionary articles, including the one propounding the (Special) Theory of Relativity.

1907 Discovers the principle of equivalence.

1910 Joins the German University in Prague.

1911 Predicts the deflection of light.

1914 Appointed professor at the University of Berlin, heads the Kaiser Wilhelm research institute for physics, and becomes a member of the Prussian Academy of Sciences.

1916 Publishes his General Theory of Relativity. Makes significant contributions to the theory of radiation and to statistical mechanics.

1919 Divorces his first wife to marry his cousin, Elsa Loewenthal-Einstein. His theories on the deflection of light are confirmed during a solar eclipse; as a result, he becomes universally known overnight.

1921 Visits the United States for the first time, together with Chaim Weizmann, to raise funds for the Hebrew University and for the purchase of land in Palestine.

1922 Publishes his first paper on unified field theory. Receives the Nobel Prize for physics for 1921. Ever more famous, reluctantly becomes a prominent public figure.

1923 Visits Palestine. Gives inaugural scientific lecture at the future site of the Hebrew University.

1925 Signs manifesto against obligatory military service.

1927 Begins debate with physicist Niels Bohr on the foundations of quantum mechanics.

1930 Becomes increasingly active on behalf of pacifism.

1932 Appointed professor at the Institute for Advanced Study at Princeton University.

1933 Declares that he will never return to Germany and emigrates to the United States, settling for good in Princeton.

1939 After learning that a secret German uranium project is in progress, signs his famous letter to President Franklin D. Roosevelt, recommending American research on nuclear weapons, which eventually results in the Manhattan Project and the development of the atomic bomb.

1945 Shocked by the extent of the Holocaust and by the nuclear bombing of Hiroshima and Nagasaki, becomes involved with philanthropic causes, particularly charitable and social organizations to help refugees from Nazi Germany.

1947 Becomes very active on behalf of disarmament and world government.

1952 Offered the presidency of Israel by Ben-Gurion.

1955 Signs the Russel-Einstein Manifesto warning of the nuclear threat.
Dies on April 18, at the age of 76.

Provenance and Description of Einstein's 1912 Manuscript on the Special Theory of Relativity

The manuscript reproduced here is the earliest surviving autograph manuscript by Einstein on the (Special) Theory of Relativity; it is the most substantial and significant of the surviving scientific manuscripts written by Einstein during his period of greatest creativity, before the end of World War I.

Date and genesis of the manuscript:

Although undated, this detailed exposition of the theory of relativity was, as will be noted more fully below, assigned to the year 1912 by Einstein himself. The text was commissioned from Einstein by Erich Marx, of the University of Leipzig, as a major contribution to the multivolume *Handbuch der Radiologie,* then in planning, over which Marx had general editorship. The *Handbuch der Radiologie,* published by the Akademische Verlagsgesellschaft, was organized to a very high standard. Apart from Einstein, at least six of its contributors were, or would become, Nobelists in physics or chemistry (Peter Debye, Max von Laue, Hendrik A. Lorentz, Sir Owen Richardson, Sir Ernest Rutherford, and Wilhelm Wien). The project, however, suffered from various delays and changes in the next years, especially because of the intervention of World War I. Einstein's contribution was originally planned for volume V, later volume VI, on theories of radiology and radioactivity. The preface to volume IV (September 1916; published 1917) still listed Einstein's paper, which had presumably been sent to the publisher already in 1912, as forthcoming.

However, the passage of years greatly altered Einstein's point of view, and by the early 1920s, as Marx began to gather the papers he had solicited on the theoretical bases of radiology and radioactivity, he found that Einstein was willing neither to let now outdated work of his be published, nor to take the time from his heavily occupied life to bring it up to date. At the beginning of 1922, Marx wrote to Einstein about the problem: the manuscript Einstein had submitted concerned only the Special Theory of Relativity, "for you had not yet worked out your General Theory." He asked whether Einstein could contribute an additional brief outline on the General Theory of Relativity, to be added to the manuscript they already had in hand. Einstein responded promptly but briefly: he begged Marx to see it as no sign of unfriendliness, but he was too busy to enlarge his manuscript. He suggested the names of several other physicists who could be approached, including Hermann Weyl, Max von Laue, and the young prodigy Wolfgang Pauli. In early March, Marx wrote again to Einstein. After long consideration, he had decided that an additional text on the General Theory of Relativity, since Einstein flatly refused to write it, would not be necessary. But the existing manuscript on the Special Theory (being much more relevant to physics at the atomic level), was critical. Would Einstein please review it again and send some sort of supplement? It would be good, he wrote, to be able to replace the manuscript's present date, from before the War, with a more current one. One may reasonably surmise that these words "das Datum von vor dem Kriege" (date from before the war), rang a loud alarm in Einstein's mind. On the back of Marx's letter, in the hand of his

secretary (his stepdaughter Ilse), is a draft reply clearly dictated by Einstein, very probably for a telegram: his manuscript, "written in the year 12," is unfortunately obsolete and therefore cannot be published at all; he will never agree to it.

Further discussion with the publishers must have taken place, in an attempt to retrieve the situation. Einstein suggested that his young assistant in Berlin, Jacob Grommer, could revise the ten-year-old paper, and on 20 March 1922 a director of the Akademische Verlagsgesellschaft, Dr. Jolowicz, wrote to Einstein asking for Grommer's address. In the same letter, Jolowicz offered to make a gift of any of his house's publications that would be useful for the newly forming Hebrew University in Jerusalem: a philanthropic work in which Einstein was actively involved. But later correspondence shows that this alternative plan still did not yield a solution. Two years later, 17 March 1924, Professor Marx wrote again to Einstein: he continued to await "Einstein & Grommer," the only article outstanding; can Einstein please persuade Grommer to bring this to a rapid conclusion?

Once more, an obstacle presented itself: now, it appears, Einstein was no longer willing to co-sign the article with Grommer. This was not, surely, from lack of respect for Grommer, for whom his regard was high and with whom he worked closely in these years, but from a reluctance to sign a piece of writing to which he had not really contributed. A final letter from Marx to Einstein of 1 July 1924 shows Marx in near desperate straits. He sent to Einstein copies of all the other contributors' articles, by such eminent friends and colleagues as Hendrik A. Lorentz, Pieter Zeeman, Peter Debye, Max von Laue, and Arnold Sommerfeld. Einstein was to note in particular how specifically Sommerfeld's contribution calls out for one by Einstein. The volume could not be further delayed, that would be unfair to the others. Marx accepted that Einstein would not co-sign with Grommer, but the publishers were very eager to have something under Einstein's own name. Could he not produce something brief but substantive, as Sommerfeld had done with regard to quantum theory—for instance, indicating how the Relativity Theory explained the changes of perihelion (of Mercury). Marx's pleas had no effect. Volume VI of his *Handbuch der Radiologie* was published in early 1925, with no contribution either from Einstein or from Grommer. In his preface, dated November 1924, Marx referred to the delays imposed by World War I, but also to the "abrupt withdrawal" of promised work by unnamed contributors ("plötzliches Zurückziehen von Mitarbeiterzusagen"): much harsher words than one is accustomed to find in editorial prefaces.

It must be noted that in the years 1922–24, Einstein's world and life were radically different from what they had been in 1912. Ten years before, in his early thirties, he was a young, newly minted full professor, well known among his peers but unknown outside their circle. By 1922, he had become a figure of world fame as the very type of scientific genius. The first experimental verifications of different aspects of his Theory of Relativity, in particular the observational confirmation by the 1919 solar eclipse expedition of his prediction of the curvature of space, conferred on him, in the public's eye, a status greater than any scientist had ever enjoyed (or suffered from) in his lifetime. At the same time, Einstein's own sense of responsibility awakened in him the obligation to involve himself in major political and humanitarian issues, including Zionism.

We may suppose that Einstein's pleas of insufficient time were not lightly made. With specific regard to scientific work, it should be recalled that in the summer of 1924, when Marx was making his final pleas for Einstein's attention, Einstein was just becoming acquainted and deeply involved with the work of the Indian physicist Satyendra Nath Bose, which led to the development of the Bose-Einstein statistics of

quantum mechanics, and the postulate of the bizarre state of matter known as Bose-Einstein condensation, created in near absolute-zero conditions and observed for the first time in 1995.

Contents of the manuscript:

The year 1905 has been often called Einstein's annus mirabilis, in parallel with Newton's plague-year annus mirabilis of 1666. In 1900, in a position paper on the state of physics, Henri Poincaré had posited three outstanding problems to which current physics had no satisfactory explanations: the apparent absence of ether drift, Brownian motion, and the photoelectric effect. Within a period of fifteen weeks in 1905, Einstein submitted to the *Annalen der Physik,* from the Swiss Patent Office in Bern, papers that revolutionized—and, implicitly, interconnected—the way in which all three problems were perceived. The last of these papers, the first on Relativity proper, has been described by its most thorough historian as "unparalleled in the history of science in its depth, breadth, and sheer intellectual virtuosity" (A. I. Miller). The "reception history" of the (Special) Theory of Relativity was complex over the following years. Einstein himself developed much more explicitly by 1907–08 the principle of mass-energy equivalence ($E=mc^2$). Other important developments were contributed by Max Planck, Max von Laue, Paul Ehrenfest, and Hermann Minkowski.

By the time Einstein's 1912 paper was solicited by Marx, relativity theory had become a familiar physical model for all members of the then small community of theoretical physicists, although among some there continued to be uncertainty of whether Einstein's theory differed in a significant way from Hendrik A. Lorentz's electrodynamics. Einstein himself had hitherto given no comprehensive overview of the consequences and implications of relativity theory; the first monograph on the subject had been Laue's *Das Relativitätsprinzip,* 1911. The present manuscript of 1912 becomes, therefore, the most detailed and comprehensive explication of the (Special) Theory of Relativity that Einstein ever composed. Being written for fellow physicists rather than as a popularization, it is conceived at a high level of scientific and mathematical sophistication, and shows a greater thoroughness in covering all aspects of relativity than is found in any of his earlier articles. The work is divided into four chief sections:

1. *Die Grundlinien der Lorentz'schen Elektrodynamik* [Fundamentals of Lorentz's electrodynamics]
2. *Elementare Darlegung der Grundlagen und wichtigste Folgerungen der Relativitätstheorie* [Elementary exposition of the foundations and most important consequences of relativity theory]
3. *Einige Begriffe und Sätze der vierdimensionalen Vektoren und Tensoren Theorie, die für das Verständnis von Minkowskis Darstellung der Relativitätstheorie nötig sind* [Several concepts and statements of the four-dimensional vector and tensor theory, necessary for understanding Minkowski's exposition of relativity theory]
4. *Elektrodynamik bewegter Körper* [Electrodynamics of moving bodies]

Einstein notes in detail at several points the surprising consequences of relativity theory and the difficulties of experimental verification of its predictions (all these predictions have been verified experimentally, though sometimes not until decades after 1912). Thus, on fos. 29–30, in his discussion of the slowing of all clocks in

velocity, he notes that this consequence has struck many physicists as so absurd — "abenteuerlich" — that they have felt compelled to reject it. However counterintuitive, it is now a scientific commonplace. He treats at length (fos. 38 sqq.) the mass-energy equivalence that is embodied in what is surely the most famous mathematical equation of this century, $E=mc^2$. His formulation of the equation has, at first, an algebraic constant added to the energy statement, which he then crossed out in draft. Almost prophetically, in light of the eventual uses made of it, Einstein refers to the mass-energy equivalence as (fo. 42) "the most important of all the consequences of relativity theory."

The lengthy exposition in section 3, on Minkowski's four-dimensional space-time continuum, is of major significance as a document of Einstein's own intellectual development. Minkowski, a Göttingen mathematician of exceptional brilliance, first presented a reworking of the Lorentz and Einstein equations into tensor form in an address to a scientific congress in Cologne in September 1908, where he boldly announced, "From this hour on, space as such and time as such must fade away into shadow, and only a kind of union of the two will maintain its reality." Four months later, he died very prematurely from appendicitis. Einstein himself was at first not greatly impressed by Minkowski's reformulation, seeing it as a mathematician's "superfluous erudition" ("überflüssiges Gelehrsamkeit": Pais, p. 152). By 1912, however, Einstein and his colleagues Sommerfeld and Ehrenfest had come to understand the great importance of Minkowski's concepts such as spacelike and timelike vector, light cone, and world line, and to incorporate this more challenging but ultimately simpler mathematical structure into their own way of thinking. The present paper is the first evidence for Einstein's use of Minkowski's tensor calculus, which was critical for Einstein's development of the General Theory of Relativity. Einstein's introduction to the 1916 offprint or monograph form of *Die Grundlage der allegemeinen Relativitätstheorie* made special acknowledgment of Minkowski.

Physical evidence of the manuscript, and Einstein in the year 1912:

One of the most fascinating and significant aspects of the present manuscript is its time-capsule quality. The year 1912 was important in both Einstein's intellectual and emotional life. Following his years in the Swiss Patent Office, Einstein found his first full academic appointment, in the fall of 1909, as associate ("extraordinary") professor at the University of Zurich. In the following year he was nominated to a full professorship at the German University in Prague, an appointment he took up in the spring of 1911. But in the course of 1911 he entered negotiations to bring him back to Zurich, as a full professor at the Institute of Technology (ETH: Eidgenössische Technische Hochschul) with a large salary. By the beginning of 1912 he had accepted this new position, planning a return to Zurich in midsummer. In the meanwhile, in April, he had traveled to Berlin for a week, where he met with Emil Warburg, Max Planck, and other leading scientists, and had first conversations about the possibility of an appointment in Berlin. On this visit, he also met, for the first time since childhood, his first cousin Elsa Einstein Lowenthal, who was now a grown woman several years his elder, divorced, with two daughters. During the week, which included an excursion to the Wannsee, they fell rapidly in love, and Einstein found an additional motive for wanting to move to Berlin. But for the shorter term he, his first wife Mileva, and their two sons moved from Prague to Zurich at the end of July.

During these months Einstein was presumably working on the present manuscript, and a strong signal of the move from Prague to Zurich rests in the combined evidence of its paper stocks and inks. The first 46 leaves are written in brown ink on two unwatermarked paper stocks (1–29; 30–46), the second of which is of very middling quality. Fos. 47 and after show a difference both of paper and ink. The higher quality paper, with a dolly-rolled laid-line pattern, is watermarked "Biberist," a Swiss paper manufacturer; and the ink changes from brown to black. It may reasonably be postulated that the first 46 leaves were written in Prague, and that fos. 47 to the end mark the move back to Zurich, and hence belong to the late summer to early autumn of 1912. This physical change corresponds closely to a change in Einstein's drafting of his third section, on Minkowski's "Vector and tensor theory." Einstein had begun this section on fo. 43 and continued it, on the "Prague" paper stock, through fo. 46. Then, on the Swiss paper of fo. 47 he extensively continued and revised this, canceling for example, more than half his text on fo. 46, and beginning it anew on fo. 47. In black ("Zurich") ink, Einstein went back to the beginning of section 3 (fo. 43) and revised its heading from "Einige Begriffe und Sätze der vierdimensionellen Geometrie . . ." to: ". . . der vierdimensionellen Vektoren und Tensoren Theorie." It is very probable that this reflects one aspect of Einstein's collaboration in Zurich with his close friend the mathematician Marcel Grossmann, former fellow student who was now dean of the mathematics and physics faculty at the ETH. By anecdote, the collaboration began when Einstein went to his friend and said, "Grossmann, you've got to help me, or else I'll go crazy!" ("Grossmann, Du musst mir helfen, sonst werd'ich verrückt!"). The collaboration was critical to Einstein's working out the much more complex mathematics needed to deal successfully with gravitational forces and, hence, to develop the General Theory of Relativity. In October of that year, Einstein wrote to Sommerfeld that he had never worked so hard in his life and had learned for the first time a true respect for mathematics as a thing of beauty: "compared to this problem, the original relativity theory was child's play" *(Collected Papers,* V, 505; see Pais ch. 12 passim).

Three other leaves of the manuscript are apparently written on yet a fourth paper stock, and these (13a, 20a, and the final leaf, 70) probably represent Einstein's latest work on the manuscript, perhaps carried out in early 1913. As the editors of the Einstein Papers have noted, on fo. 20a, Einstein referred to research by the Dutch scientist Willem de Sitter on astronomical evidence for the constance of the velocity of light. De Sitter's paper was published in the *Physikalische Zeitschrift* on 1 May 1913. However, Einstein's draft gives de Sitter's name in the aurally corrupted form "Pexider," from which it may be inferred that he had only heard about, not read, de Sitter's researches.

Publication history of the manuscript and its value as historical evidence:

Every page of Einstein's manuscript shows extensive reworking. Quite regularly, two stages or strata, original composition and redrafting, can be distinguished; but very often, three or even four different strata can be extracted, each revealing the changes of thought of a great scientist who was also a notably sensitive stylist and expositor. Einstein had, moreover, the pleasant habit of striking out his canceled words very lightly, with a single line or a few hatchings. Rarely, if ever, is his original formulation illegible. Probably no other work from Einstein's hand reveals more closely the process of his expository thoughts. This manuscript has recently been published by the

editors of The Collected Papers of Albert Einstein, as part of volume 4, *The Swiss Years: Writings, 1912–1914* (September 1995).

Rarity of Einstein scientific manuscripts:

Early scientific manuscripts of Einstein are of the highest rarity. Professor John Stachel has written (Einstein, *Collected Papers,* I, xxvii–xxviii): "Einstein made no systematic attempt to preserve his papers before about 1920. Prior to that time, he routinely discarded manuscripts of published articles and very few have been preserved. . . . The earliest evidence of Einstein's concern for the fate of his papers dates from 1921, when he agreed to give his correspondence to the Prussian State Library for its collection of scientists' papers." Only in 1928, when his devoted secretary Helen Dukas joined him, were careful steps taken to gather and maintain the Einstein Archives, which are now at the Hebrew University, Jerusalem. None of the original manuscripts survive for the articles of the "annus mirabilis" 1905. The combined power of Einstein's name, his handwriting, and his patriotism is embodied in one of the most curious episodes in auction history. In late 1943, Einstein wrote out a fair copy of the first relativity paper, "Zur Elektrodynamik bewegter Körper," which he contributed to the U.S. war bond campaign. On Thursday, 3 February 1944, a war bond auction was held, Clifton Fadiman presiding, in Kansas City, Missouri, at which this fair copy was "sold" (or better, awarded) to the Kansas City Life Insurance Company in return for its investment of $6,500,000 in war bonds. Later in the year, the manuscript was donated to the Library of Congress, where it remains. Of Einstein's early scientific writings, excluding a few course-lecture notebooks in the Einstein Archives and letters, only the following autographs seem to be known:

1895?: Über die Untersuchung des Aetherzustandes in magnetischen Felde. A.MS, 5_ pp. A "sample" scientific paper written at the age of 16 for Einstein's uncle, Caesar Koch. Published in: *The Collected Papers of Albert Einstein:* Volume 1 (The Early Years, 1879-1902), 1987, pp. 6-9. [location: private collection; sold Christie's London, 24 June 1987, lot 170]

1912: The present paper [(Spezielle) Relativitätstheorie]. A.MS, 72 p. Descended in the family of the intended publisher; sold Sotheby's New York, 2 December 1987, lot 53. Published in: *The Collected Papers of Albert Einstein:* Volume 4 (The Swiss Years, 1912–1914), 1995, pp. 9–108. [location: The Israel Museum, Jerusalem].

1913: Untitled study written jointly by Einstein and Michele Besso, on the precession of the perihelion of Mercury. A.MS, (approximately half in Einstein's hand, half in Besso's), 51 p. Published in *The Collected Papers of Albert Einstein:* Volume 4 (The Swiss Years, 1912–1914), 1995, pp. 360–473, 630–82. [location: Besso Family Trust, Geneva].

1914: Eine Methode zur statistischen Verwertung von Beobachtungen scheinbar unregelmässig quasiperiodisch verlaufender Vorgänge. A.MS, 4 pp. Notes for a paper on statistical interpretation of scientific data, delivered to the Swiss Society of Physics, conference meeting of 28 February 1914, Basel. Published in: *The Collected Papers of Albert Einstein:* Volume 4 (The Swiss Years, 1912–1914), 1995, pp. 603–7. A French translation of the paper was published in: *Archives des sciences physiques et naturelles,* 4th ser. 37 (1914), pp. 254–56: Méthode pour la détermination de valeurs statistiques d'observations concernant des grandeurs soumises à des fluctuations irrégulières. (not in Weil, etc., not mentioned in *The Collected Papers of Albert Einstein).* [location: Einstein Archives, Hebrew University, Jerusalem].

1914: Antrittsrede [to the Prussian Academy of Sciences, 2 July]. A.MS, 2 pp. The text of Einstein's speech on induction into the Prussian Academy of Sciences, published in: Preussische Akademie der Wissenschaften, *Sitzungsberichte,* 1914, II, pp. 739–42 (Weil 67A). [location: Einstein Archives, Hebrew University, Jerusalem].

1914: Zum Relativitätsproblem. A.MS, 11 pp. The text of Einstein's article in: *Scientia* 15 (1914) 337–48. (Weil 69) [location: private collection].

1916: Die Grundlagen der allgemeinen Relativitätstheorie. A.MS, 45 pp. The fundamental paper on the General Theory of Relativity. Published in: *Annalen der Physik,* 4th ser., 49 (1916), pp. 769–822. (Weil 80, 80A). [location: Schwadron Collection, Hebrew University, Jerusalem].

1919–20: Grundgedanken und Methoden der Relativitätstheorie in ihrer Entwicklung dargestellt. A.MS., 35 pp. Unpublished. This paper was commissioned from Einstein by the English journal *Nature* in late 1919, for a projected special issue on Relativity, in follow-up to the 6 November report to the Royal Society of the British solar eclipse expedition verifying the gravitational curvature of space. The paper turned out to be too long for the journal, whose eventual "Relativity" issue of 17 February 1921 contained only a brief translated precis. (Weil 119). [location: The Pierpont Morgan Library, Dannie and Hettie Heineman Collection (see Pierpont Morgan Library, Report to the Fellows 1975–1978, pp. 30, 221)].

EINSTEIN'S 1912 MANUSCRIPT ON THE SPECIAL THEORY OF RELATIVITY

$$\mathfrak{E} = \frac{mc^2}{\sqrt{1-\frac{q^2}{c^2}}}$$

SECTION ONE. [1912—1914] [1]

AN OUTLINE OF LORENTZ'S ELECTRODYNAMICS

§1. The Fundamental Maxwell-Lorentz Equations for the Vacuum in the Absence of Electrically and Magnetically Polarizable Bodies

A complete understanding of the justification of the theory that we today designate as the theory of relativity is possible only if we call to mind the outlines of the development of electrodynamics since Maxwell. We will therefore briefly review the basic ideas of this development.

Quantity of electricity. If electrically charged corpuscles are (continually) present, then, as we know, their electrical charges $e_1\ e_2\ldots$ can be defined as follows: Let the ratio $e_1{:}e_1$ etc. be equal to the ratio of the forces experienced by the corpuscles in the *same* electrostatic field.[2] This definition is possible because the above ratio is known from experience to be independent of the choice of the field. Further, one can determine the absolute magnitude of the charges e by postulating that the repulsive force exerted by two corpuscles (say, those with indices 1 and 2) on each other in vacuum is equal to $\frac{e_1 e_2}{4\pi r^2}$. This expression contains, on the one hand, Coulomb's empirical law and, on the other hand, the definition of the electrostatic unit as introduced by Heaviside[3] <and H. A. Lorentz>. Thus, $\frac{e}{\sqrt{4\pi}}$ is the quantity of electricity measured in the customary electrostatic units.

Electrical field strength. By the electrical field strength[4] $\mathfrak{e}$ in vacuum one understands the vector that—when multiplied by e—is equal to the vector of the force that the field exerts on a stationary corpuscle endowed with charge e.

The magnetic pole strength p and the magnetic field strength in vacuum $\mathfrak{h}$ shall be defined in an analogous way.

The first system of Maxwell's equations in the absence of electrically and magnetically polarizable media. A temporally constant, closed electrical circuit produces a magnetic field that is determined by the following rule: The line integral of the magnetic field strength over an arbitrary closed curve (line element $d\mathfrak{s}$) is equal to the surface integral of the vector $\mathfrak{i}$ of the electrical current density divided by a certain constant c. The components of this vector are thereby defined as those electrical quantities that in unit time pass through surfaces of magnitude 1 that are normal to the coordinate axes. Thus, with an appropriate choice of the orientation of the surface normal,* we have the equation

$$\int \mathfrak{h}\, d\mathfrak{s} = \frac{1}{c}\int \mathfrak{i}_n\, d\sigma \qquad \ldots (1)$$

Since, according to Stokes's theorem

$$\int \mathfrak{h}\, d\mathfrak{s} = \int (\operatorname{curl} \mathfrak{h})_n\, d\sigma,$$

and since the above equation should be valid for arbitrary curves and thus also for plane curves of infinitesimally small dimensions, there follows from it

$$\operatorname{curl} \mathfrak{h} = \frac{1}{c}\mathfrak{i}. \qquad \ldots (1a)$$

The normal should point left for an observer who is passing along the curve in the specified direction with his head forward, and is looking toward the interior of the surface.

~~1. Kapitel.~~ 1. Abschnitt. Die Grundlinien der Lorentz'schen Elektrodynamik. (1)

§1.

Die Maxwell-Lorentz'schen Grundgleichungen (beim Fehlen elektrisch oder magnetisch polarisierbarer Körper) ~~für das Vakuum.~~

Die Berechtigung derjenigen Theorie, welche wir heute als Relativitätstheorie bezeichnen, kann wohl nur dann ganz erkannt werden, wenn man sich die Entwicklung der Elektrodynamik seit Maxwell in ihren Grundlinien vergegenwärtigt. Wir wollen deshalb die Grundgedanken dieser Entwicklung kurz an uns vorbeiziehen lassen.

<u>Elektrizitätsmenge.</u> Liegen (unendlich) kleine elektrisch geladene Körperchen vor, so können wir die elektrischen Ladungen e_1, e_2 … derselben bekanntlich so definieren. Das Verhältnis $e_1 : e_2$ etc. sei gleich dem Verhältnis der Kräfte, welche die Körperchen in <u>demselben</u> elektrostatischen Felde erfahren. Diese Definition ist möglich, weil das genannte Verhältnis erfahrungsgemäss unabhängig ist von der Wahl des Feldes. Es lässt sich ferner die absolute Grösse der Ladungen e durch die Festsetzung bestimmen, dass die abstossende Kraft, welche zwei Körperchen (etwa diejenigen mit den Indizes 1 und 2) ~~aufeinander~~ im Vakuum aufeinander ausüben, gleich $\frac{e_1 e_2}{4\pi r^2}$ sei. Dieser Ausdruck (~~Festsetzung~~) enthält einerseits das Coulomb'sche Erfahrungsgesetz, andererseits diejenige Festsetzung der elektrostatischen Einheit, wie sie durch Heaviside ~~und H. A. Lorentz~~ eingeführt wurde. Es ist also $\frac{e}{\sqrt{4\pi}}$ die in gewöhnlichem elektrostatischem Masse gemessene Elektrizitätsmenge.

<u>Elektrische Feldstärke.</u> Unter elektrischer Feldstärke ~~in~~ ~~einem~~ versteht man jenen Vektor, welcher mit e multipliziert gleich ist dem Vektor der Kraft, welche ~~in dem Felde~~ auf ein ruhendes Körperchen von der Ladung e vom Felde ausgeübt wird.

In analoger Weise sei die magnetische Polstärke μ und die magnetische Feldstärke $\mathfrak{h}$ im Vakuum definiert.

<u>Erstes Maxwell'sches Gleichungssystem, falls elektrisch und magnetisch polarisierbare Medien fehlen.</u> ~~In ruhenden geschlossenen Leitern~~ ~~erzeugt die~~ ~~Änderung~~ Zeitlich konstanter, geschlossener elektrischer Ströme (Kreis) erzeugen ein Magnetfeld, welches durch folgende Regel bestimmt ist. Das Linienintegral der magnetischen Feldstärke über eine beliebige geschlossene Kurve (Linienelement $d\mathfrak{s}$) ist gleich dem Flächenintegral des ~~elektrischen Stromvektors~~ Vektors $\mathfrak{i}$ der elektrischen Stromdichte, mit einer gewissen Konstanten c dividiert. Die Komponenten dieses Vektors sind dabei definiert als diejenigen elektrischen Mengen, welche pro Zeiteinheit Flächen von der Grösse 1 passieren, die auf den Koordinatenachsen senkrecht stehen. Es gilt also bei geeigneter Wahl des Sinnes der Flächennormale* die Gleichung

$$\int \mathfrak{h}\, d\mathfrak{s} = \frac{1}{c} \int \mathfrak{i}_n \, d\sigma \quad \ldots \ldots (1)$$

Da nach dem Stokes'schen Satze

$$\int \mathfrak{h}\, d\mathfrak{s} = \int (\operatorname{curl} \mathfrak{h})_n \, d\sigma$$

ist, und die obige Gleichung ~~auch~~ für beliebige Kurven, also auch für (ebene) Kurven von unendlich kleinen Abmessungen gültig sein soll, so folgt aus ihr

$$\operatorname{curl} \mathfrak{h} = \frac{1}{c}\, \mathfrak{i} \quad \ldots \ldots (1a)$$

---

* Die Normale soll nach links zeigen für einen Beobachter, der die Kurve in der festgesetzten Richtung ~~mit~~ mit dem Kopfe voran durchläuft und nach dem Inneren der Fläche schaut.

But equations (1) and (1a) can claim general validity only in the case where the current is stationary. For if one takes the divergence on both sides of (1a), one obtains div $\mathfrak{i}$ = 0; this equation cannot be generally valid since there are also currents that are not closed. Maxwell got rid of this contradiction by introducing the hypothesis that, besides the conduction current $\mathfrak{i}$, "the electrical displacement current $\dot{\mathfrak{e}}$ also participates in the production of the magnetic field.[5] The equation thus completed reads*

$$\text{curl } \mathfrak{h} = \frac{1}{c}(\dot{\mathfrak{e}} + \mathfrak{i}). \qquad \ldots (1b)$$

Taking the divergence again on both sides, one obtains

$$0 = \frac{\partial}{\partial t}(\text{div } \mathfrak{e}) + \text{div } \mathfrak{i}.$$

The following should be noted about this equation. In the definition of current density the law of the conservation of the quantity of electricity was already implicity assumed. For the quantity of electricity traversing a cross section cannot be measured directly, but only the change that the electrical charge on a body undergoes with time. Implicitly, we set this equal to the quantity of electricity that flows from the body, or flows to it; i.e., in order to invest our definition of $\mathfrak{i}$ with physical meaning, we already had to assume the indestructibility of the electrical quantities. The last-derived equation shows, therefore, that div $\mathfrak{e}$ is nothing other but the density ρ of the electrical charge. Hence we can set

$$\text{div } \mathfrak{e} = \rho \qquad \ldots (2)$$

It can also easily be shown that this equation agrees with the definition of the unit of electrical quantity given earlier.

Equation (1b) is exactly correct—at least as far as our current knowledge goes—as long as the material carriers of the currents and charges are at rest. But if moving, electrically charged bodies are present, curl $\mathfrak{h}$ is different from zero even if $\dot{\mathfrak{e}}$ and $\mathfrak{i}$ vanish. Rowland has shown that a rotating, electrically charged, metal disk generates a magnetic field.[6] From this experiment one can conclude that an electrically charged body moving with velocity $\mathfrak{q}$ and charge density ρ is equivalent to a current distribution of current density $\rho\mathfrak{q}$. This current, which is called a convection current, has been introduced into the theory by Maxwell.[7] Therefore, if one aims for an equation that is valid for all processes in vacuum, one will replace (1b) with the equation

$$\text{curl } \mathfrak{h} = \frac{1}{c}(\dot{\mathfrak{e}} + \mathfrak{i} + p\mathfrak{q}). \qquad \ldots (1c)$$

But we can write this equation in a simpler form if we make use of a hypothesis by the application of which H. A. Lorentz has tremendously advanced electrodynamics. The law of the conservation of the quantities of electricity suggests the assumption that changes of location are the only changes that the quantities of electricity can experience, or in other words, that electrical currents are always

* "$\dot{\mathfrak{e}}$" denotes the time derivative of $\mathfrak{e}$; i.e., the vector with the components $\frac{\partial \mathfrak{e}_x}{\partial t}, \frac{\partial \mathfrak{e}_y}{\partial t}, \frac{\partial \mathfrak{e}_z}{\partial t}$.

Die Gleichung (1) bezw. (1a) kann aber nur in dem Falle allgemeine Gültigkeit beanspruchen, dass ~~die Ströme~~ der Strom ein stationärer ist. Denn nimmt man auf beiden Seiten von (1a) die Divergenz, so ergibt sich $\mathrm{div}\, i = 0$; diese Gleichung kann nicht allgemein gelten, da es ungeschlossene Ströme gibt. Diesen Widerspruch beseitigte Maxwell, indem er die Hypothese einführte, dass neben dem Leitungsstrom i auch der „elektrische Verschiebungsstrom" $\dot{e}$ sich an der Erzeugung des Magnetfeldes beteilige. Die demgemäss ergänzte Gleichung lautet*

$$\mathrm{curl}\, f = \frac{1}{c}(\dot{e} + i) \ldots\ldots (1b)$$

Nimmt man wieder auf beiden Seiten die Divergenz, so erhält man

$$0 = \frac{\partial}{\partial t}(\mathrm{div}\, e) + \mathrm{div}\, i.$$

Zu dieser Gleichung ist Folgendes zu bemerken. Bei der Definition der Stromdichte (i) wurde bereits implizite der Satz von der Erhaltung der Elektrizitätsmengen vorausgesetzt. Denn es lässt sich die einen Querschnitt passierende Elektrizitätsmenge nicht direkt messen, sondern nur die Änderung, welche die ~~Elektrizität~~ elektrische Ladung eines Körpers mit der Zeit erfährt. Diese setzten wir implizite gleich der Elektrizitätsmenge, welche von dem Körper wegfliesst, bezw. dem Körper zufliesst, d. h. wir mussten, um unserer Definition von i einen physikalischen Sinn zu geben, bereits die Unzerstörbarkeit der elektrischen Mengen voraussetzen. Die zuletzt abgeleitete Gleichung zeigt daher, dass $\mathrm{div}\, e$ nichts anderes ist als die Dichte ρ der elektrischen Ladung. Es ist also zu setzen

$$\mathrm{div}\, e = \rho \ldots\ldots (2)$$

Es lässt sich auch leicht zeigen, dass diese Gleichung mit der ~~la~~ oben gegebenen Festsetzung über die Einheit der elektrischen Menge im Einklang ist.

Die Gleichung (1b) ist ~~aber immer~~ – soweit unser heutiges Wissen reicht – exakt zutreffend, solange die materiellen Träger der Ströme und Ladungen in Ruhe sind. In dem Falle aber, dass bewegte, elektrisch geladene Körper vorhanden sind, ist $\mathrm{curl}\, f$ auch dann von null verschieden wenn $\dot{e}$ und i verschwinden. Rowland zeigte, dass eine sich drehende elektrisch geladene Metallscheibe ein Magnetfeld erzeugt. Aus diesem Versuche lässt sich folgern, dass ein mit der Geschwindigkeit ~~*~~ q bewegter elektrisch geladener Körper von der Ladungsdichte ρ einer ~~[illegible]~~ Stromverteilung von der Stromdichte ρq äquivalent ist. Dieser Strom, welchen man Kon~~Ver~~vektionsstrom ~~schiebungs~~ nennt, wurde bereits von Maxwell in die Theorie eingeführt. Soll daher eine für alle Vorgänge im Vakuum gültige Gleichung angestrebt werden, so wird man (1b) durch die Gleichung

$$\mathrm{curl}\, f = \frac{1}{c}(\dot{e} + i + \rho q) \ldots\ldots (1c)$$

ersetzen.

~~Nehmen wir in dieser Gleichung auf beiden Seiten die Divergenz~~

Diese Gleichung können wir aber einfacher schreiben wenn wir uns ~~auf den Boden~~ einer Hypothese bedienen durch deren Durchführung H. A. Lorentz die Elektrodynamik ungemein gefördert hat. Der Satz von der Erhaltung der Elektrizitätsmengen legt die Annahme ~~es~~ nahe, dass die einzigen Änderungen, welche Elektrizitätsmengen erfahren können, ~~durch Bewegung~~ Ortsänderungen seien, dass also mit anderen Worten elektrische Ströme stets Konvektionsströme seien. Nach dieser

* Mit „$\dot{e}$" ist die zeitliche Ableitung des Vektors e bezeichnet, d. h. der Vektor mit den Komponenten $\frac{\partial e_x}{\partial t}, \frac{\partial e_y}{\partial t}, \frac{\partial e_z}{\partial t}$.

convection currents. According to this assumption, one has to imagine that an apparently charge-free conductor contains positively and negatively charged corpuscles the sum of whose charges is zero; in that case, the electrical current is due to a motion of the positively charged corpuscles with respect to the negatively charged ones. On this conception, equation (1c) reduces to

$$\text{curl}\ \mathfrak{h} = \frac{1}{c}\,(\dot{\mathfrak{e}} + p\mathfrak{q}). \qquad \ldots \text{(1d)}$$

The second system of Maxwell's equations in the absence of electrically and magnetically polarizable media. As we know, Maxwell's second system of equations is the expression of Faraday's law of electromagnetic induction for electric circuits at rest and for infinitely small spaces, if one also includes the hypothesis that the magnetoelectrically induced electromotive force is essentially the same as an electrical field strength. For slowly changing magnetic fields the law of induction is by the equation

$$\int \mathfrak{e}\,d\mathfrak{s} = -\frac{1}{c}\int \dot{\mathfrak{h}}_n d\sigma, \qquad \ldots \text{(3)}$$

where the integral on the left side is to be extended over an arbitrary closed curve, and the one on the right over an arbitrary surface that has the above curve as its boundary. The orientation of the normal in the surface integral is connected with the direction in which the line integral is taken by the same rule as above.

If one applies the equation to plane surfaces of infinitesimal extension in both dimensions, one obtains the vector equation that holds, according to Maxwell, for arbitrarily fast processes [8]

$$\text{curl}\ \mathfrak{E} = -\frac{1}{c}\dot{\mathfrak{h}} \qquad \ldots \text{(3a)}$$

If one performs the "div" operation on both sides, one obtains

$$\frac{\partial}{\partial t}(\text{div}\ \mathfrak{h}) = 0.$$

Thus, if div $\mathfrak{h}$ were at all different from zero, it would have to be temporally constant. Since this is out of the question from a physical standpoint, one has to set

$$\text{div}\ \mathfrak{h} = 0. \qquad \ldots \text{(4)}$$

Thus, to sum up, we have the following. If one confines oneself to electrically and magnetically nonpolarizable bodies and if one assumes with H. A. Lorentz that there do not exist electrical currents of any kind other than convection currents, then the electromagnetic equations read

$$\begin{aligned} &\text{curl}\ \mathfrak{h} = \frac{1}{c}(\dot{\mathfrak{e}} + \mathfrak{q}\rho) \qquad &&\text{curl}\ \mathfrak{e} = -\frac{1}{c}\dot{\mathfrak{h}} \\ &\text{div}\ \mathfrak{e} = \rho &&\text{div}\ \mathfrak{h} = 0 \end{aligned} \qquad \text{(I)}$$

Annahme hat man sich in einem dem Anscheine nach ladungsfreien Leiter positiv und negativ elektrisch geladener Körperchen von der Ladungssumme null zu denken; der elektrische Strom beruht dann auf einer Bewegung der ~~positiv~~ geladenen gegenüber den negativ geladenen Körperchen. Bei dieser Auffassung reduziert sich Gleichung (1c) auf

$$\mathrm{curl}\, \mathfrak{h} = \frac{1}{c}(\dot{\mathfrak{e}} + \rho\, \mathfrak{q}) \quad \ldots\ldots (1d).$$

Zweites Maxwell'sches Gleichungssystem, falls elektrisch und magnetisch polarisierbare Medien fehlen. Maxwells zweites Gleichungssystem ist bekanntlich der Ausdruck des Faraday-schen magnetelektrischen Induktionsgesetzes (für ruhende Stromkreise und für ~~das~~ unendlich kleine Räume), wenn man noch die Hypothese hinzunimmt, dass die magnetelektrisch induzierte elektromotorische Kraft dem Wesen nach gleich sei einer elektrischen Feldstärke. Für langsam veränderliche Magnetfelder ~~drückt sich~~ ist das Induktionsgesetz durch die Gleichung

$$\int \mathfrak{e}\, ds = -\frac{1}{c}\int \dot{\mathfrak{h}}_n\, d\sigma \quad \ldots\ldots (3)$$

wobei das Integral der linken Seite über eine beliebige geschlossene Kurve, das der rechten Seite über eine beliebige Fläche zu erstrecken ist, die jene Kurve zur Randlinie hat. Die Richtung der Normale beim Flächenintegral ist mit dem Durchlaufungssinn des Linienintegrales durch die nämliche Regel ~~verbunden~~ verknüpft wie oben.

Wendet man die Gleichung auf ebene Flächen von (in beiden Dimensionen) ~~beiderseits~~ unendlich kleiner Ausdehnung an, so erhält man die nach Maxwell für beliebig rasch verlaufende Vorgänge gültige Vektorgleichung

$$\mathrm{curl}\, \mathfrak{e} = -\frac{1}{c}\dot{\mathfrak{h}} \quad \ldots\ldots (3a)$$

Führt man auf beiden Seiten die Operation „div" aus, so erhält man

$$\frac{\partial}{\partial t}(\mathrm{div}\, \mathfrak{h}) = 0.$$

Wenn also $\mathrm{div}\, \mathfrak{h}$ überhaupt von Null verschieden wäre, so müsste es zeitlich unveränderlich sein. Da dies vom physikalischen Standpunkt aus ausgeschlossen ist, hat man ~~gleichzeitig~~

$$\mathrm{div}\, \mathfrak{h} = 0 \qquad (4)$$

zu setzen.

Zusammenfassend ergibt sich also Folgendes. Beschränkt man sich auf elektrisch und magnetisch nicht polarisierbare Körper, und nimmt man mit H. A. Lorentz an, dass es keine andersartigen elektrischen Ströme als Konvektionsströme gibt, so lauten die elektromagnetischen Gleichungen

$$\left.\begin{aligned} \mathrm{curl}\, \mathfrak{h} &= \frac{1}{c}(\dot{\mathfrak{e}} + \mathfrak{q}\rho) \qquad & \mathrm{curl}\, \mathfrak{e} &= -\frac{1}{c}\dot{\mathfrak{h}} \\ \mathrm{div}\, \mathfrak{e} &= \rho & \mathrm{div}\, \mathfrak{h} &= 0 \end{aligned}\right\} (I)$$

§2. *Energy and Momentum in Lorentz's Electrodynamics in the Absence of Electrically and Magnetically Polarizable Bodies*

The energy principle. If one takes the scalar product of the first of equations (I) by $c\mathfrak{e}$ and the third of equations (I) by $-c\mathfrak{h}$, and sums, one obtains, <using the rule of calculation ( )> [9] after simple transformation, the equation [10]

$$-\text{div}\, c\,[\mathfrak{e},\mathfrak{h}] = \frac{\partial}{\partial}\left(\frac{\mathfrak{e}^2 + \mathfrak{h}^2}{2}\right) + \rho\mathfrak{q}\mathfrak{e} \qquad \ldots (5)$$

or, if one integrates over an arbitrary volume while applying ( )

$$\frac{\partial}{\partial t}\left\{\int \frac{1}{2}(\mathfrak{e}^2 + \mathfrak{h}^2)\,d\tau\right\} = \int c[\mathfrak{e}\mathfrak{h}]_n\,d\sigma - \int \rho\mathfrak{q}\mathfrak{e}\,d\tau, \qquad \ldots (5a)$$

if one denotes the component along the inside normal of the boundary surface of the space under consideration by $[\mathfrak{e}, \mathfrak{h}]_n$. One may conceive of this equation as the equation for the energy balance of the electromagnetic field. In that case $w = \frac{1}{2}(\mathfrak{e}^2 + \mathfrak{h}^2)$ is to be conceived as the density of the electromagnetic energy and $\mathfrak{s} = c[\mathfrak{e}, \mathfrak{h}]$ as the vector of the energy flow, while $\rho\mathfrak{q}\mathfrak{e}$ signifies the energy taken from the electromagnetic field per unit volume and time.

The momentum conservation law. If one takes the vector product of the first of equations (I) by $\mathfrak{h}$ and the third by $-\mathfrak{e}$, and then sums, one obtains first

$$[\text{curl}\,\mathfrak{h},\mathfrak{h}] + [\text{curl}\,\mathfrak{e},\mathfrak{e}] = \frac{\partial c}{\partial x}\left\{\frac{1}{c}[\mathfrak{e},\mathfrak{h}]\right\} + \left[\frac{\mathfrak{q}\rho}{c}, \mathfrak{h}\right]$$

Integrating this by parts, one obtains easily the identities

$$-[\text{curl}\,\mathfrak{h},\mathfrak{h}]_x = \frac{\partial}{\partial x}\left(\frac{\mathfrak{h}^2}{2} - \mathfrak{h}_x\mathfrak{h}_x\right) - \frac{\partial}{\partial y}(\mathfrak{h}_x\mathfrak{h}_y) - \frac{\partial}{\partial z}(\mathfrak{h}_y\mathfrak{h}_z)$$

$$-[\text{curl}\,\mathfrak{e},\mathfrak{e}]_x = \mathfrak{e}_x\,\text{div}\,\mathfrak{e} + \frac{\partial}{\partial x}\left(\frac{\mathfrak{e}^2}{2} - \mathfrak{e}_x\mathfrak{e}_x\right) - \frac{\partial}{\partial y}(\mathfrak{e}_x\mathfrak{e}_y) - \frac{\partial}{\partial z}(\mathfrak{e}_y\mathfrak{e}_z).$$

Thus, if one sets [11]

$$P_{xx} = \frac{1}{2}(\mathfrak{e}^2 + \mathfrak{h}^2) - \mathfrak{e}_x\mathfrak{e}_x - \mathfrak{h}_x\mathfrak{h}_x \mid p_{xy} = p_{yx} = -\mathfrak{e}_x\mathfrak{e}_y - \mathfrak{h}_x\mathfrak{h}_y \mid$$

$$p_{xz} = p_{zx} = -\mathfrak{e}_x\mathfrak{e}_z - \mathfrak{h}_x\mathfrak{h}_z$$

etc. and takes into account the second of equations (I), one obtains three relations, the first of which is

$$\rho\left\{\mathfrak{e} + \left[\frac{\mathfrak{q}}{c}, \mathfrak{h}\right]\right\}_x = -\frac{\partial p_{xx}}{\partial x} - \frac{\partial p_{xy}}{\partial x} - \frac{\partial p_{xz}}{\partial x} - \frac{1}{c^2}\frac{\partial s_x}{\partial t} \qquad \ldots (6)$$

If we denote the left-hand side of this equation, which is the x component of a vector, by $\mathfrak{f}_x$ and integrate both sides over a finite volume, we obtain

$$\int \mathfrak{f}_x\,d\tau = -\frac{\partial}{\partial t}\left\{\int \frac{1}{c^2}s_x\,d\tau\right\} + \int (p_{xx}\cos(nx) + p_{xy}\cos(ny) + p_{xz}\cos(nz))\,d\sigma$$

or in short

$$\int \mathfrak{f}_x\,d\tau = -\frac{\partial}{\partial t}\left\{\int \frac{1}{c^2}s_x\,d\tau\right\} + \int p_{nx}\,d\sigma \qquad \ldots (6a).$$

§ 2.

Energie und Impuls in der Lorentz'schen Elektrodynamik beim Fehlen elektrisch und magnetisch polarisierbarer Körper.

Energieprinzip. Multipliziert man die erste ~~gl~~ der Gleichungen (I) skalar mit $c\mathfrak{e}$, die dritte der Gleichungen (I) mit $-c\mathfrak{h}$ und addiert, so erhält man ~~bei Benutzung der Rechnungsregel ( )~~ nach einfacher Umformung die Gleichung:

$$-\operatorname{div} c[\mathfrak{e},\mathfrak{h}] = \frac{\partial}{\partial t}\left(\frac{\mathfrak{e}^2+\mathfrak{h}^2}{2}\right) + \varrho\,\mathfrak{v}\,\mathfrak{e} \quad \ldots\ldots (5)$$

oder, indem man über ein beliebiges Volumen integriert unter Anwendung von ( )

$$\frac{\partial}{\partial t}\left\{\int \frac{1}{2}(\mathfrak{e}^2+\mathfrak{h}^2)\,d\tau\right\} = \int c[\mathfrak{e}\,\mathfrak{h}]_n\,d\sigma - \int \varrho\,\mathfrak{v}\,\mathfrak{e}\,d\tau, \quad \ldots (5a)$$

falls man mit $[\mathfrak{e},\mathfrak{h}]_n$ die Komponente nach der inneren Normale der Begrenzungsfläche des betrachteten Raumes bezeichnet. Man kann diese Gleichung als die Gleichung der Energiebilanz des elektromagnetischen Feldes auffassen. Dann ist $W = \frac{1}{2}(\mathfrak{e}^2+\mathfrak{h}^2)$ als die Dichte der elektromagnetischen Energie, $\mathfrak{S} = c[\mathfrak{e},\mathfrak{h}]$ als Vektor der Energieströmung aufzufassen, während $\varrho\,\mathfrak{v}\,\mathfrak{e}$ die pro Volumen und Zeiteinheit dem elektromagnetischen Felde entzogene Energie bedeutet.

Impulssatz. Multipliziert man die erste der Gleichungen (I) vektoriell mit $\mathfrak{h}$, die dritte mit $-\mathfrak{e}$ und addiert dann, so erhält man zunächst

$$[\operatorname{curl}\mathfrak{h},\mathfrak{h}] + [\operatorname{curl}\mathfrak{e},\mathfrak{e}] = \frac{\partial}{\partial t}\left\{\frac{1}{c}[\mathfrak{e},\mathfrak{h}]\right\} + \left[\frac{\mathfrak{v}\varrho}{c},\mathfrak{h}\right]$$

Es folgt nun leicht durch partielle Integration ~~bei Anwendung der zweiten der Gleichungen (I)~~ die Identitäten

~~$[\operatorname{curl}\mathfrak{h},\mathfrak{h}] = \frac{\partial(\mathfrak{h}_x\mathfrak{h}_x)}{\partial x} - \partial\mathfrak{e}_x\mathfrak{e}_y$~~

$$-[\operatorname{curl}\mathfrak{h},\mathfrak{h}]_x = \frac{\partial}{\partial x}\left(\frac{\mathfrak{h}^2}{2} - \mathfrak{h}_x\mathfrak{h}_x\right) - \frac{\partial}{\partial y}(\mathfrak{h}_x\mathfrak{h}_y) - \frac{\partial}{\partial z}(\mathfrak{h}_x\mathfrak{h}_z)$$

$$-[\operatorname{curl}\mathfrak{e},\mathfrak{e}]_x = \mathfrak{e}_x\operatorname{div}\mathfrak{e} + \frac{\partial}{\partial x}\left(\frac{\mathfrak{e}^2}{2} - \mathfrak{e}_x\mathfrak{e}_x\right) - \frac{\partial}{\partial y}(\mathfrak{e}_x\mathfrak{e}_y) - \frac{\partial}{\partial z}(\mathfrak{e}_x\mathfrak{e}_z).$$

Setzt man also

$$p_{xx} = \frac{1}{2}(\mathfrak{e}^2+\mathfrak{h}^2) - \mathfrak{e}_x\mathfrak{e}_x - \mathfrak{h}_x\mathfrak{h}_x \quad\Big|\quad p_{xy} = p_{yx} = -\mathfrak{e}_x\mathfrak{e}_y - \mathfrak{h}_x\mathfrak{h}_y \quad\Big|\quad p_{xz} = p_{zx} = -\mathfrak{e}_x\mathfrak{e}_z - \mathfrak{h}_x\mathfrak{h}_z$$

etc., so erhält man unter Berücksichtigung der zweiten der Gleichungen (I) drei Beziehungen, deren erste lautet:

$$\varrho\left\{\mathfrak{e} + \left[\frac{\mathfrak{v}}{c},\mathfrak{h}\right]\right\}_x = -\frac{\partial p_{xx}}{\partial x} - \frac{\partial p_{xy}}{\partial y} - \frac{\partial p_{xz}}{\partial z} - \frac{1}{c^2}\frac{\partial \mathfrak{S}_x}{\partial t} \quad \ldots (6)$$

Bezeichnen wir die linke Seite dieser Gleichung, welche die x Komponente eines Vektors ist, mit $\mathfrak{f}_x$, und integrieren wir beiderseits über ein endliches Volumen, so erhalten wir

$$\int \mathfrak{f}_x\,d\tau = -\frac{\partial}{\partial t}\left\{\int \frac{1}{c^2}\mathfrak{S}_x\,d\tau\right\} + \int\left(p_{xx}\cos(nx) + p_{xy}\cos(ny) + p_{xz}\cos(nz)\right)d\sigma$$

oder kürzer $\int \mathfrak{f}_x\,d\tau = -\frac{\partial}{\partial t}\left\{\int \frac{1}{c^2}\mathfrak{S}_x\,d\tau\right\} + \int p_{nx}\,d\sigma \quad \ldots\ldots (6a).$

H. A. Lorentz designates the vector $\mathfrak{f}$ as the ponderomotive [12] force per unit volume exerted by the field on the electricity.[1] By means of this conception he does justice to the Biot-Savart empirical law, on the one hand, and, on the other hand, he succeeds in having his electrodynamics satisfy the energy law and the momentum law. For if, in setting up (5a) and (6a), one chooses the integration space in such a manner that the field strengths vanish permanently at the boundaries of the space, then when the relation

$$\mathfrak{q}\mathfrak{f} = \mathfrak{q}\rho\{\mathfrak{e} + \left[\frac{\mathfrak{q}}{c}, \mathfrak{h}\right]\} = \rho\mathfrak{q}\mathfrak{e}$$

is taken into account, these relations turn into

$$\int \mathfrak{q}\mathfrak{f}d\tau = -\frac{\partial}{\partial t}\{\int w d\tau\} \qquad \text{. . . (5b)}$$

and

$$\int \mathfrak{f}d\tau = -\frac{\partial}{\partial t}\{\int \frac{1}{c^2} s d\tau\} \qquad \text{. . . (6b)}$$

It follows from (5b) that the work $\int \mathfrak{q}\mathfrak{f}d\tau$ that is performed by the field per unit time is always associated with an equally large decrease of the quantity $\int w\, d\tau$ of the field; hence we have to view the latter as the energy of the total field, and we remain in accord with the energy law if we view w as the density of the energy of the electromagnetic field. It follows from (6b) that the volume integral of the ponderomotive force exerted by the total field is equal to the decrease of the vector integral $\int \frac{1}{c^2}\mathfrak{s}d\tau$; hence, we have to view the latter as the expression for the momentum of the total field, and we remain in accord with the momentum law if we view $\frac{1}{c^2}\mathfrak{s}$ as the (vectorial) momentum density of the field.

From the formal point of view, we note that the momentum and energy laws are satisfied in Lorentz's theory due to the existence of equations of the following form for the ponderomotive force $\mathfrak{f}$ per unit volume:

$$\left.\begin{array}{l} \mathfrak{f}_x = -\frac{\partial p_{xx}}{\partial x} - \frac{\partial p_{xy}}{\partial y} - \frac{\partial p_{xz}}{\partial z} - \frac{1}{c^2}\frac{\partial \mathfrak{s}_x}{\partial t} \\ - - - - - - - - - - - - - - - - \\ - - - - - - - - - - - - - - - - \\ \mathfrak{q}\mathfrak{f} = -\frac{\partial \mathfrak{s}_x}{\partial x} - \frac{\partial \mathfrak{s}_y}{\partial y} - \frac{\partial \mathfrak{s}_z}{\partial z} - \frac{\partial w}{\partial t} \end{array}\right\} \qquad (7)$$

It should be emphasized that H. A. Lorentz's expression for the force density $\mathfrak{f}$ is one of the empirically best supported results in electrodynamics. It yields immediately the electromotive forces that are induced in conductors moving in stationary magnetic fields,[13] the simplest case of the Zeemann effect,[14] the way magnetic fields influence the cathode rays.[15] But we will not get into this here.

---

[1] The force $\mathfrak{K}$ acting on an infinitesimally small charged particle of volume V is obviously

$$\mathfrak{K} = \mathfrak{f}V = \rho V\{\mathfrak{e} + \left[\frac{\mathfrak{q}}{c}, \mathfrak{h}\right]\} = \varepsilon\{\mathfrak{e} + \left[\frac{\mathfrak{q}}{c}, \mathfrak{h}\right]\},$$

where ε denotes the electrical charge.

H. A. Lorentz bezeichnet nun den Vektor $\mathfrak{f}$ als die pro Volumeneinheit ponderomotorische Kraft, welche vom Felde auf die Elektrizität ausgeübt wird.[1] Durch diese Auffassung wird er einerseits den Biot-Savart-schen Erfahrungsgesetze gerecht, andererseits erreicht er es, dass seine Elektrodynamik dem Energiesatz und dem Impulssatz Genüge leistet. Denn wählt man für die Aufstellung von (5a) und (6a) den Integrationsraum derart, dass an den Grenzen des Raumes die Feldstärken dauernd verschwinden, so gehen diese Gleichungen bei Berücksichtigung der Relation

$$\mathfrak{v}\mathfrak{f} = \mathfrak{v}\varrho\left\{\mathfrak{e} + \left[\frac{\mathfrak{v}}{c}, \mathfrak{h}\right]\right\} = \varrho\mathfrak{v}\mathfrak{e}$$

über in

$$\int \mathfrak{v}\mathfrak{f}\,d\tau = -\frac{\partial}{\partial t}\left\{\int W\,d\tau\right\} \quad \ldots\ldots (5b)$$

und

$$\int \mathfrak{f}\,d\tau = -\frac{\partial}{\partial t}\left\{\int \frac{1}{c^2}\mathfrak{s}\,d\tau\right\} \quad \ldots\ldots (6b)$$

Aus (5b) folgt, dass mit der Arbeitsleistung $\int \mathfrak{v}\mathfrak{f}\,d\tau$ stets eine gleichgrosse Abnahme der Grösse $\int W\,d\tau$ des Feldes pro Zeiteinheit verbunden ist; die letztere haben wir daher als die Energie des gesamten Feldes anzusehen, und wir bleiben im Einklang mit dem Energiesatze, wenn wir W als die Dichte der Energie des elektromagnetischen Feldes ansehen. Aus (6b) folgt, dass das Volumintegral der vom gesamten Felde ausgeübten ponderomotorischen Kraft gleich ist der Abnahme des vektoriellen Integrales $\int \frac{1}{c^2}\mathfrak{s}\,d\tau$; das letztere haben wir daher als den Ausdruck des Impulses des gesamten Feldes anzusehen, und wir bleiben im Einklang mit dem Impulssatze, wenn wir $\frac{1}{c^2}\mathfrak{s}$ als die (vektorielle) Impulsdichte des Feldes ansehen.

In formaler Beziehung bemerken wir, dass der Impulssatz und Energiesatz in der Lorentz'schen Theorie dadurch erfüllt ist, dass für die ponderomotorische Kraft $\mathfrak{f}$ pro Volumeneinheit Gleichungen von der Form existieren:

$$\left.\begin{array}{l} \mathfrak{f}_x = -\dfrac{\partial p_{xx}}{\partial x} - \dfrac{\partial p_{xy}}{\partial y} - \dfrac{\partial p_{xz}}{\partial z} - \dfrac{1}{c^2}\dfrac{\partial \mathfrak{s}_x}{\partial t} \\ - - - - - - - - - - - - - \\ - - - - - - - - - - - - - \\ \mathfrak{v}\mathfrak{f} = -\dfrac{\partial \mathfrak{s}_x}{\partial x} - \dfrac{\partial \mathfrak{s}_y}{\partial y} - \dfrac{\partial \mathfrak{s}_z}{\partial z} - \dfrac{\partial W}{\partial t} \end{array}\right\} (7)$$

Es ist zu betonen, dass der H. A. Lorentz'sche Ausdruck für die Kraftdichte $\mathfrak{f}$ zu den durch die Erfahrung am besten fundierten Ergebnissen der Elektrodynamik gehört. Aus ihm ergeben sich unmittelbar die elektromotorischen Kräfte, die in in ruhenden Magnetfeldern bewegten Leitern induziert werden, der einfachste Fall des Zeemaneffektes, die Art der Einwirkung von Magnetfeldern auf Kathodenstrahlen. Es soll jedoch hierauf nicht näher eingegangen werden.

---

1) Die auf ein unendlich kleines geladenes Teilchen vom Volumen V wirkende Kraft $\mathfrak{K}$ ist offenbar

$$\mathfrak{K} = \mathfrak{f}V = \varrho V\left\{\mathfrak{e} + \left[\frac{\mathfrak{v}}{c}, \mathfrak{h}\right]\right\} = \varepsilon\left\{\mathfrak{e} + \left[\frac{\mathfrak{v}}{c}, \mathfrak{h}\right]\right\},$$

wobei ε die elektrische Ladung des Teilchens bezeichnet.

§3. Completion of Lorentz's Theory for the Case Where Electrically and Magnetically Polarizable Media Are Present (Bodies at Rest)

The brilliant idea through which H. A. Lorentz advanced electrodynamics and optics in such an extraordinary way can be expressed as follows. Every influence of matter on the electromagnetic field and, conversely, every influence of the electromagnetic field on matter is based on the fact that matter contains movable electrical masses that interact with the electromagnetic field according to equations (I). In this connection, H. A. Lorentz conceives of electricity as being bound to corpuscles of molecular dimensions (electrons in the broader sense), a conception whose validity is hardly doubted today.[16] But complications are thereby created for the theory, in that one is dealing here with field quantities that vary rapidly with location and that are to be replaced, then, by suitable mean values.[17] One can avoid these complications without doing any essential damage if one proceeds in the following way.

According to the picture that Lorentz's conception gives, we have to conceive an electrically polarizable body in the following way. In every unit volume of a body in an electrically neutral state there are present at least two approximately evenly distributed kinds of electrons of zero total charge. But these are not freely movable; instead, they are linked to matter by elastic forces (in the simplest case). An electric field displaces the positive and negative electrons from their equilibrium position by means of oppositely directed forces. In this process, the electromagnetic field varies extremely rapidly with location. We avoid this by conceiving of the positive as well as the negative electrons of the same kind as being combined into continua. In the simplest case, we have to picture an inertia-free electrical continuum of positive density and, in the case of an electrically unexcited body, one of equally great negative density, linked elastically to the matter. If we also wish to represent the conductivity of the body, we introduce, in addition, two further electrically opposite density-continua that can move relative to the body by overcoming a kind of friction. There is nothing strange in the introduction of several continua at the same location if one realizes that this is only an idealization aimed at avoiding mathematical complications. In this way we make it possible for the field strengths $\mathfrak{e}$ and $\mathfrak{h}$ to retain their simple meaning. Likewise, equations (I) retain their general validity, the only difference being that in the first of these equations we have to put $\sum \mathfrak{q}\rho$ in place of $\mathfrak{q}\rho$, where the sum is to be extended over all continua; likewise, in the second equation ρ has to be replaced by $\sum \rho$.

Polarization. Let ρ_v be the density of an electrical continuum bound to matter (polarization density). Let the continuum be displaced infinitesimally relative to the matter. Let the vector of this displacement be $\mathfrak{q}_v'$. In that case, $\rho_v \mathfrak{q}_v'$ is also a vector, and we will designate the sum $\sum \rho_v \mathfrak{q}_v'$, which also has the character of a vector, as the vector $\mathfrak{p}$ of the electrical polarization. Then $\mathfrak{p}_x$ is the sum of the electrical

§ 3.

Vervollständigung der Lorentz'schen Theorie für den Fall, dass elektrische und
magnetisch polarisierbare Medien vorhanden sind. (Ruhende Körper).

Der geniale Gedanke, durch den H. A. Lorentz die Elektrodynamik
und Optik so ungemein förderte lässt sich so ausdrücken. Jede Beeinflussung des elektromagnetischen Feldes durch Materie und umgekehrt
jede Beeinflussung der Materie durch das elektromagnetische Feld beruht
darauf, dass die Materie bewegbare elektrische Massen enthält, die nach
den Gleichungen (I) mit dem elektromagnetischen Felde in Wechselwir-
kung stehen. Hiebei denkt sich H. A. Lorentz ~~mit guten~~ die Elektrizität an molekular kleine Korpuskeln (Elektronen im weiteren Sinne)
gebunden, eine Auffassung, ~~deren~~ [Berechtigung] heute wohl kaum mehr in Zweifel
gezogen wird. Hiedurch werden aber für die Theorie Komplikationen
erzeugt, indem man es mit örtlich rasch variierenden Feldgrössen zu thun
hat, die dann durch geeignete Mittelwerte zu ersetzen sind. Man kann
diese Komplikationen, ohne der Sache wesentlich Abbruch zu thun, in
folgender Weise vermeiden.

Nach dem durch Lorentz' Auffassung gegebenen Bilde haben wir
uns einen elektrisch polarisierbaren ~~und gleichzeitig elektr~~ Körper wie folgt
zu denken. In jeder Volumeinheit [des Körpers] sind im elektrisch neutralen Zustande
[wenigstens zwei] ~~g~~ annähernd gleichmässig verteilte Elektronen[arten] von der Ladungssumme
null vorhanden. Diese sind aber nicht frei beweglich, sondern durch
(im einfachsten Falle) elastische Kräfte an ~~einen Art~~ die Materie gekettet.
Ein elektrisches Feld verschiebt die positiven und negativen Elektronen
durch entgegengesetzt gerichtete Kräfte aus ihrer Gleichgewichtslage.
Das elektromagnetische Feld ist hiebei ~~und~~ örtlich ungeheuer rasch
veränderlich~~es~~. Dies vermeiden wir, indem wir sowohl die positiven
wie die negativen Elektronen gleicher Art zu Kontinuen zusammen-
gefasst denken. In dem einfachsten Falle ~~dass die Dispersion nicht~~
~~berücksichtigt wird,~~ haben wir uns ~~trägheit~~ [bei elektrisch nicht erregtem Körper] ein [trägheitsfreies] elektrisches Kontinuum ~~a~~ [von]
positiver Dichte und eines von gleich grosser negativer Dichte als
elastisch an die Materie gekettet vorzustellen. Wollen wir auch noch
die Leitfähigkeit des Körpers darstellen, so führen wir [ausserdem noch] ~~auch~~ zwei
[entgegengesetzt] elektrische Dichte-Kontinua ein, die relativ zum Körper unter
Überwindung einer Art Reibung beweglich sind. Die Einführung
mehrerer Kontinua an demselben Orte hat nichts Befremdliches,
wenn man sich vergegenwärtigt, dass es sich hier nur um eine
Idealisierung zu Vermeidung mathematischer Komplikationen
handelt. Auf diese Weise erreichen wir es, dass die Feldstärken e und h ihre
einfache Bedeutung behalten. Ebenso behalten die Gleichungen (I) ihre
allgemeine ~~Bedeutung~~ [Gültigkeit] mit dem einzigen Unterschiede, dass wir in
deren erste statt $\mathfrak{v}\varrho$ zu setzen haben $\Sigma\,\mathfrak{v}\varrho$, wobei die Summe über alle
Kontinua zu erstrecken ist; ebenso ist in der zweiten Gleichung ϱ durch
$\Sigma\varrho$ zu ersetzen.

Polarisation. Sei ϱ_ν die Dichte eines an die Materie gebundenen elektrischen
Kontinuums. (Polarisationsdichte) Dieses ~~erfahrt~~ sei relativ zur Materie unendlich wenig verschoben.
Der Vektor dieser Verschiebung sei $\mathfrak{v}_\nu'$. Dann ist $\varrho_\nu\,\mathfrak{v}_\nu'$ ebenfalls ein Vektor,
und wir wollen die Summe $\Sigma\,\varrho_\nu\,\mathfrak{v}_\nu'$, welche ebenfalls Vektorcharakter hat,
als den Vektor ~~der~~ $\mathfrak{P}$ der elektrischen Polarisation bezeichnen. Es ist dann $\mathfrak{P}_x$
die Summe der elektrischen Mengen, welche beim Herstellen der „elektrischen Polarisation"

quantities that traverse, per unit surface, a surface perpendicular to the X-axis, per unit surface during the establishment of "electrical polarization"; the projection $\mathfrak{p}_n$ on the normal to an arbitrarily oriented surface has an analogous meaning. If a dielectric borders on a vacuum, and if $\mathfrak{n}$ denotes the normal to the surface of the latter directed toward the latter, then $\mathfrak{p}_n$—taken on the surface—therefore measures an electrical surface density (surface density of the bound electricity) that is produced by the polarization densities on the surface of the dielectric.

Dielectric displacement Dielectric constant. By "dielectric displacement" $\mathfrak{d}$ we understand the sum vector sum of the electrical field strength $\mathfrak{e}$ and the polarization $\mathfrak{p}$. Thus, we set

$$\mathfrak{d} = \mathfrak{e} + \mathfrak{p}.$$

According to experience, in many cases one should set $\mathfrak{d} = \varepsilon\mathfrak{e}$ for isotropic dielectrics at rest, where ε denotes a constant characteristic of the dielectric, the "dielectric constant." From the equations that define $\mathfrak{d}$ and ε we obtain

$$\mathfrak{p} = (\varepsilon - 1)\mathfrak{e},$$

which equation likewise claims validity only for dielectrics *at rest.*

Physical justification of Lorentz's conception of the nature of the dielectric. A homogeneous body consisting of a dielectric, nonconductive substance that becomes electrically polarized possesses on its surface, according to the above, a layer of (bound) electricity ($\mathfrak{p}_n$ per surface unit), while the entire density of (bound) electricity vanishes on the inside of the body. If the body is set into motion, then the electrical quantities on the surface are set into motion as well. Thus, according to the first of equations (I), they must produce a magnetic field and, if no change in the state of polarization of the particles of the body takes place during the motion, the magnetic field so excited will be the only cause of the excitation of the magnetic field. The existence of this magnetic field has been proved by Röntgen and Eichenwald.[18] Furthermore, we have seen in §2 that moving electrical quantities in an electromagnetic field are acted upon by forces that are given by the expression $\mathfrak{e} + \left[\frac{\mathfrak{q}}{c}, \mathfrak{h}\right]$—referred to a unit quantity of electricity. If the dielectrica really owe their electrifiability to bound quantities of electricity, then that force must act on the latter as well; hence, if the arrangement is such that no electrical field is present, then the bound electrical masses of a dielectric moving in a magnetic field must be acted upon by forces in exactly the same way as if the dielectric were at rest and an electrical field strength of the magnitude $\left[\frac{\mathfrak{q}}{c}, \mathfrak{h}\right]$ acted upon it. A polarization of the magnitude $\mathfrak{p} = (\varepsilon - 1)\left[\frac{\mathfrak{q}}{c}, \mathfrak{h}\right]$ will therefore occur in the dielectric. The existence of such a polarization has been proved by Wilson.[19]

Electromagnetic equations for bodies at rest. We start with the second of equations (I). If we denote the individual densities of the "bound" electricity by ρ_g and the single density of the conduction electricity by ρ_l, then, in accordance with the explanations given at the beginning of this §, we must set

$$\operatorname{div} \mathfrak{e} = \sum \rho_g + \sum \rho_l.$$

But from the definition given for $\mathfrak{p}$ there follows immediately

$$\rho_g = -\operatorname{div} \mathfrak{p}_g$$

$$\sum \rho_g = -\operatorname{div} \mathfrak{p}.$$

Furthermore, if, for the sake of brevity, we denote the total density of the conduction electricity $\sum \rho_l$ by ρ, then our equation becomes

pro Flächeneinheit

pro Flächeneinheit durch eine senkrecht zur X-Achse gelegte Fläche hindurchtritt; eine
analoge Bedeutung hat die Normalprojektion $\mathfrak{P}_n$ zu einer beliebig orientierten
Fläche. Grenzt ein Dielektrikum an das Vakuum, und bezeichnet n die nach dem
letzteren hin gerichtete Normale der Oberfläche des letzteren, so misst also $\mathfrak{P}_n$ ~~die~~
~~Fläch~~ – an der Oberfläche genommen – eine elektrische Flächendichte (Flächendichte
der gebundenen Elektrizität) welche die Polarisationsdichten an der Oberfläche
des Dielektrikums liefern.

Dielektrizitätskonstante

Dielektrische Verschiebung. Unter der „dielektrischen Verschiebung" ~~verst~~ $\mathfrak{d}$ verstehen
wir die Summe Vektorsumme der elektrischen Feldstärke $\mathfrak{e}$ und der Polarisation $\mathfrak{P}$. Es ist also

$$\mathfrak{d} = \mathfrak{e} + \mathfrak{P}$$

in vielen Fällen

gesetzt. Für ruhende ~~Diele~~ isotrope Dielektrika ist nach der Erfahrung $\mathfrak{d} = \varepsilon \mathfrak{e}$
zu setzen, wobei ε eine für das Dielektrikum charakteristische Konstante,
die „Dielektrizitätskonstante" bedeutet. Aus den Definitionsgleichungen
für $\mathfrak{d}$ und ε folgt

$$\mathfrak{P} = (\varepsilon - 1)\,\mathfrak{e},$$

welche Gleichung ebenfalls nur für ruhende Dielektrika Gültigkeit beansprucht.

Physikalische Berechtigung der Lorentz'schen Auffassung von der Natur des
Dielektrikums. Eine homogener Körper aus dielektrischer, nichtleitender
~~Substanz~~, welcher ~~homogen~~ elektrisch polarisiert wird, besitzt an seiner Oberfläche nach
dem Vorigen eine Belegung (gebundener) Elektrizität ($\mathfrak{P}_n$ pro Flächeneinheit)
während im Inneren des Körpers die gesamte Dichte der (gebundenen) Elektrizität
verschwindet. Wird der Körper bewegt, so werden die elektrischen Mengen an der
Oberfläche ebenfalls bewegt. Sie müssen also nach der ersten der Gleichungen (I) ein
Magnetfeld erzeugen, und zwar wird das so erregte Magnetfeld, falls eine Änderung
des Polarisationszustandes der Teilchen des Körpers während der Bewegung nicht
erfolgt, die einzige ~~Ma~~ Ursache der magnetischen Felderregung sein. Die Existenz dieses Magnetfeldes wurde von Röntgen und Eichenwald nachgewiesen.
Wir haben ferner in §2 gesehen, dass auf ~~in einem~~ bewegte elektrische Mengen
im elektromagnetischen Felde Kräfte wirken, welche – auf die Einheit
der Elektrizitätsmenge bezogen – durch den Ausdruck $\mathfrak{e} + \left[\frac{\mathfrak{q}}{c}, \mathfrak{h}\right]$ gegeben sind.
Falls die Dielektrika wirklich ihre Elektrisierbarkeit gebundenen Elektrizitätsmengen verdanken, so muss jene Kraft auch auf diese wirken; ist
die Anordnung derart getroffen, dass ein elektrisches Feld nicht vorhanden
ist, so müssen daher in einem Magnetfeld auf die ~~el~~ gebundenen elektrischen Mengen eines bewegten Dielektrikums Kräfte wirken, genau so wie wenn ~~in einem ruhenden~~ das Dielektrikum in Ruhe wäre, und eine elektrische Feldstärke vom
Betrage $\left[\frac{\mathfrak{q}}{c}, \mathfrak{h}\right]$ auf dasselbe einwirkte. Es wird daher im Dielektrikum
eine Polarisation vom Betrage $\mathfrak{P} = (\varepsilon - 1)\left[\frac{\mathfrak{q}}{c}, \mathfrak{h}\right]$ entstehen. Die Existenz
einer solchen Polarisation wurde von Wilson nachgewiesen.

Elektromagnetische Gleichungen für ruhende Körper. Wir beginnen mit der zweiten
der Gleichungen (I). Bezeichnen wir mit ϱ_g die Einzeldichten der „gebundenen" Elektrizität, mit ϱ_e die einzelne Dichte der Leitungselektrizität, so müssen wir
nach den ~~Erklärungen~~ am Anfange dieses § gegebenen Erklärungen setzen

$$\operatorname{div} \mathfrak{e} = \Sigma \varrho_g + \Sigma \varrho_e$$

Aus der für $\mathfrak{P}$ gegebenen Definition folgt aber unmittelbar

$$\varrho_g = -\operatorname{div} \mathfrak{P}_g$$

$$\Sigma \varrho_g = -\operatorname{div} \mathfrak{P}.$$

Bezeichnen wir ferner die Gesamtdichte der Leitungselektrizität $\Sigma \varrho_e$ kurz
mit ϱ, so geht unsere Gleichung über in

$$\text{div}\ \mathfrak{e} = -\text{div}\ \mathfrak{p} + \rho,$$

or—in accordance with the definitional equation given for $\mathfrak{d}$—

$$\text{div}\ \mathfrak{d} = \rho,$$

where, however, ρ denotes the density of the conduction electricity alone.

Let us now turn to the first of equations (I). The latter remains valid, but with the difference that we have to put the sum $\sum \mathfrak{q}_g \rho_g + \sum \mathfrak{q}_l \rho_l$ in place of $\mathfrak{q}\rho$. But according to our definitional equation for $\mathfrak{p}$

$$\frac{\partial \mathfrak{p}}{\partial t} = \frac{\partial}{\partial t}\left(\sum \mathfrak{q}_g' \rho_g\right) = \sum \mathfrak{q}_g \rho_g .$$

The last equation is correct because only the time derivative of the displacements $\mathfrak{q}_g'$, but not the time derivative of the ρ_g, appears to be multiplied by finite factors, and because, obviously, $\mathfrak{q}_g'$ must be replaced by $\mathfrak{q}_g$. If, in addition, we set $\sum \mathfrak{q}_l \rho_l$ equal to the total conduction current $\mathfrak{i}$, then the first of equations (I) assumes the form

$$\text{curl}\ \mathfrak{h}\ \frac{1}{c}\{\dot{\mathfrak{e}} + \dot{\mathfrak{p}} + \mathfrak{i}\}.$$

or also

$$\text{curl}\ \mathfrak{h} = \frac{1}{c}\{\dot{\mathfrak{d}} + \mathfrak{i}\}$$

The assumption of the existence of polarization electricity has no influence on the form of the third and fourth of equations (I).

Now we have to fit equations (I) to the case where magnetically polarizable bodies are present. H. A. Lorentz does this by conceiving of certain electricities as being endowed with cyclical motions; [20] from the standpoint of the pure electron theory this is also the only way that is justified. But for the sake of simplicity, we will base ourselves here on the knowledge that, as regards spatio-temporal interrelations, the magnetic polarization is a state wholly analogous to the polarization of dielectrics. Thus, we permit ourselves to conceive of magnetically polarizable bodies as being endowed with bound magnetic volume densities. In addition, one has to take into account the fact that a magnetic phenomenon that would correspond to electrical conduction does not exist. For that reason, if we denote the vector of magnetic polarization by $\mathfrak{m}$, we must supplement the last two of equations (I) by the terms $-\frac{1}{c}\ \mathfrak{m}$ and $-\text{div}\ \mathfrak{m}$, respectively. Hence, in place of (I), we obtain finally the equation

$$\left.\begin{aligned} &\text{curl}\ \mathfrak{h} = \frac{1}{c}\ (\dot{\mathfrak{e}} + \dot{\mathfrak{p}} + \mathfrak{i}) \qquad &&\text{curl}\ \mathfrak{e} = -\frac{1}{c}\ (\dot{\mathfrak{h}} + \mathfrak{m}) \\ &\text{div}\ \mathfrak{d} = -\text{div}\ \mathfrak{p} + \rho &&\text{div}\ \mathfrak{h} = -\text{div}\ \mathfrak{m} \end{aligned}\right\} \qquad \text{(Ia)}$$

or, if we introduce the vector $\mathfrak{d}$ of electric displacement as well as the vector of magnetic induction that is defined by the equation $\mathfrak{b} = \mathfrak{h} + \mathfrak{m}$, even more simply,

$$\left.\begin{aligned} &\text{curl}\ \mathfrak{h} = \frac{1}{c}\ (\dot{\mathfrak{d}} + \mathfrak{i}) \qquad &&\text{curl}\ \mathfrak{e} = -\frac{1}{c}\ \dot{\mathfrak{b}} \\ &\text{div}\ \mathfrak{d} = \rho &&\text{div}\ \mathfrak{b} = 0 \end{aligned}\right\} \qquad \text{(Ia')}$$

To these equations there are still to be added the equations that show the manner in which the vectors $\mathfrak{d}$, $\mathfrak{b}$ and $\mathfrak{i}$ depend on the field strengths $\mathfrak{e}$ and $\mathfrak{h}$. In the simplest case, we have for isotropic bodies:

$$\left.\begin{aligned} \mathfrak{d} &= \varepsilon\mathfrak{e} \\ \mathfrak{b} &= \mu\mathfrak{h} \\ \mathfrak{i} &= \lambda\mathfrak{e}. \end{aligned}\right\} \qquad (8)$$

$$\operatorname{div} \mathfrak{e} = -\operatorname{div} \mathfrak{p} + \rho,$$

oder – gemäss der für $\mathfrak{d}$ gegebenen Definitionsgleichung –

$$\operatorname{div} \mathfrak{d} = \rho,$$

wobei aber ρ die Dichte der Leitungselektrizität allein bezeichnet.

Nun ~~transformieren wir auch die~~ wenden wir uns zur ersten der Gleichungen (I). Dieselbe bleibt bestehen, mit dem Unterschiede, dass wir statt ~~ρ~~ $\eta\rho$ die Summe $\sum \eta_g \rho_g + \sum \eta_l \rho_l$ zu setzen haben. Nun ist aber gemäss unserer Definitionsgleichung für $\mathfrak{p}$

$$\frac{\partial \mathfrak{p}}{\partial t} = \frac{\partial}{\partial t}\left(\sum \eta'_g \rho_g\right) = \sum \eta_g \rho_g.$$

Die letzte Gleichung ist richtig, weil nur die zeitliche Ableitung der Verschiebungen η'_g, nicht aber die zeitliche Ableitung der ρ_g mit endlichen Faktoren multipliziert erscheinen, und weil offenbar $\dot{\eta}'_g$ durch η_g zu ersetzen ist. Setzen wir noch $\sum \eta_l \rho_l$ gleich dem gesamten Leitungsstrom $\mathfrak{i}$, so nimmt die erste der Gleichungen (I) die Form an

$$\operatorname{curl} \mathfrak{h} = \frac{1}{c}\left\{\dot{\mathfrak{e}} + \dot{\mathfrak{p}} + \mathfrak{i}\right\}$$

oder auch

$$\operatorname{curl} \mathfrak{h} = \frac{1}{c}\left\{\dot{\mathfrak{d}} + \mathfrak{i}\right\}$$

Auf ~~die~~ die Form der dritten und vierten der Gleichungen (I) hat die Annahme der Existenz von Polarisationselektrizität keinen Einfluss.

Wir haben nun die ~~dritt~~ Gleichungen (I) noch dem Falle anzupassen, dass magnetisch polarisierbare Körper vorhanden sind. H. A. Lorentz thut dies, indem er gewissen Elektrizitäten zyklische Bewegungen erteilt denkt; dieser Weg ist auch vom Standpunkte der reinen Elektronentheorie der einzig berechtigte. Wir wollen uns aber hier der Einfachheit halber auf die Erkenntnis stützen, dass inbezug auf die räumlich-zeitlichen Wechselbeziehungen die magnetische Polarisation ein der ~~elektrischen~~ die Polarisation der Dielektrika ganz analoger Zustand ist. Wir gestatten uns also die magnetisch polarisierbaren Körper als mit gebundenen magnetischen Raumdichten ausgestattet zu denken. Zudem ist zu berücksichtigen, dass ein der elektrischen Leitung entsprechendes magnetisches Phänomen nicht existiert. Wir haben deshalb, wenn wir den Vektor der magnetischen Polarisation mit $\mathfrak{m}$ bezeichnen, die letzten beiden der Gleichungen (I) durch die Glieder $-\frac{1}{c}\dot{\mathfrak{m}}$ bezw $-\operatorname{div} \mathfrak{m}$ zu ergänzen. Wir erhalten also endlich anstelle von (I) die Gleichungen:

$$\left.\begin{aligned} \operatorname{curl} \mathfrak{h} &= \frac{1}{c}(\dot{\mathfrak{e}} + \dot{\mathfrak{p}} + \mathfrak{i}) & \operatorname{curl} \mathfrak{e} &= -\frac{1}{c}(\dot{\mathfrak{h}} + \dot{\mathfrak{m}}) \\ \operatorname{div} \mathfrak{e} &= -\operatorname{div} \mathfrak{p} + \rho & \operatorname{div} \mathfrak{h} &= -\operatorname{div} \mathfrak{m} \end{aligned}\right\} \text{(Ia)}$$

oder, indem wir den Vektor $\mathfrak{d}$ der elektrischen Verschiebung sowie den durch die Gleichung $\mathfrak{b} = \mathfrak{h} + \mathfrak{m}$ definierten Vektor der magnetischen Induktion einführen, noch einfacher

$$\left.\begin{aligned} \operatorname{curl} \mathfrak{h} &= \frac{1}{c}(\dot{\mathfrak{d}} + \mathfrak{i}) & \operatorname{curl} \mathfrak{e} &= -\frac{1}{c}\dot{\mathfrak{b}} \\ \operatorname{div} \mathfrak{d} &= \rho & \operatorname{div} \mathfrak{b} &= 0 \end{aligned}\right\} \text{(I'a)}$$

(vor $\operatorname{div} \mathfrak{b} = 0$ gestrichen: ~~div $\mathfrak{h}$~~)

Zu diesen Gleichungen kommen noch jene Gleichungen hinzu, welche angeben, wie die Vektoren $\mathfrak{d}$, $\mathfrak{b}$ und $\mathfrak{i}$ von den Feldstärken $\mathfrak{e}$ und $\mathfrak{h}$ abhängen. Im einfachsten Falle ist für isotrope Körper zu setzen:

$$\left.\begin{aligned} \mathfrak{d} &= \varepsilon \mathfrak{e} \\ \mathfrak{b} &= \mu \mathfrak{h} \\ \mathfrak{i} &= \lambda \mathfrak{e}. \end{aligned}\right\} \quad (8)$$

or

$$\left.\begin{aligned} \mathfrak{p} &= (\varepsilon - 1)\,\mathfrak{e} \\ \mathfrak{m} &= (\mu - 1)\,\mathfrak{h} \\ \mathfrak{i} &= \lambda \mathfrak{e} \end{aligned}\right\}, \qquad (8')$$

where ε (the dielectric constant), μ (permeability), and λ (conductivity) are characteristic constants of the material.

Energy principle. If one multiplies the first of equations (Ia′) by $c\mathfrak{e}$ and the third by $c\mathfrak{h}$, and adds the two, one obtains, if one assumes the validity of equations (8),

$$\mathfrak{e}\mathfrak{i} = -\operatorname{div}\,\mathfrak{s} - \frac{\partial}{\partial t}\left(\frac{\mathfrak{e}\mathfrak{d} + \mathfrak{h}\mathfrak{b}}{2}\right), \qquad \ldots (9)$$

if one again sets

$$\mathfrak{s} = c[\mathfrak{e}, \mathfrak{h}].$$

If one integrates over a finite volume, one obtains, as in §2,

$$\frac{\partial}{\partial t}\left\{\int \frac{\mathfrak{e}\mathfrak{d} + \mathfrak{h}\mathfrak{b}}{2}\,d\tau\right\} = \int \mathfrak{s}_n d\sigma - \int \mathfrak{e}\mathfrak{i}\,d\tau$$

This equation expresses the energy principle for the case considered here. One sees that the vector of the energy flux depends only on the field strengths, exactly as in the case where there are no polarizable bodies, and that one has to view $\frac{\mathfrak{e}\mathfrak{d}}{2}$ as the density of the electrical energy and $\frac{\mathfrak{h}\mathfrak{b}}{2}$ as the density of the magnetic energy.

We can decompose this energy density of the electromagnetic field into the following components w, w_e, and w_m:

$$w = \frac{\mathfrak{e}^2 + \mathfrak{h}^2}{2}$$

$$w_e = \frac{1}{\varepsilon - 1}\,\frac{\mathfrak{p}^2}{2}$$

$$w_m = \frac{1}{\mu - 1}\,\frac{\mathfrak{m}^2}{2}.$$

This decomposition is called for from the physical standpoint in the light of Lorentz's theory. For w is the purely electromagnetic energy density that the field would possess even in the absence of polarizable bodies. w_e is the density of the energy that one has to apply so as to endow the medium with the polarization $\mathfrak{p}$ opposing the elastic forces between the bound electricities and matter. Thus, this energy attaches to matter and has nothing to do directly with the electromagnetic field; it need not be viewed as electromagnetic energy, being, instead, only connected with it by virtue of the properties of matter.* The same applies to w_m. We can write for (9):

$$\mathfrak{e}\mathfrak{i} = -\operatorname{div}\,\mathfrak{s} - \frac{\partial w}{\partial t} - \frac{\partial w_e}{\partial t} - \frac{\partial w_m}{\partial t}. \qquad \ldots (9a)$$

Ponderomotive forces exerted on bodies at rest. We first seek these forces for the special case where only an electrostatic field is present. We inquire after the work A' that the field supplies in the course of infinitely small displacement $\mathfrak{s}$ of the material particle. The ponderomotive forces acting on matter can then be deduced from this work. From the energy principle one obtains

$$A' = -\delta\left\{\int \frac{\mathfrak{e}\mathfrak{d}}{2}\,d\tau\right\},$$

where "δ" denotes the change that the quantity appearing behind δ experiences as

* The decomposition introduced is particularly justified on account of the case where equations (8) are to be replaced with more complicated relations. Then w retains its form, while the expressions for w_e and w_m have to be replaced by more complicated ones.

oder

$$\left.\begin{aligned} \mathfrak{p} &= (\varepsilon-1)\,\mathfrak{e} \\ \mathfrak{m} &= (\mu-1)\,\mathfrak{h} \\ \mathfrak{i} &= \lambda\,\mathfrak{e} \end{aligned}\right\} (8'),$$

wobei ε (Dielektrizitätskonstante), μ (Permeabilität) und λ (Leitvermögen) charakteristische Konstante der Materie sind.

<u>Energieprinzip.</u> Multipliziert man die erste der Gleichungen (I_a') mit $c\mathfrak{e}$, die dritte mit $c\mathfrak{h}$ und addiert beide ~~und integriert über einen begrenzten Raum~~, so erhält man unter Voraussetzung der Gültigkeit der Gleichungen (8)

$$\mathfrak{e}\mathfrak{i} = -\operatorname{div}\mathfrak{S} - \frac{\partial}{\partial t}\left(\frac{\mathfrak{e}\mathfrak{d} + \mathfrak{h}\mathfrak{b}}{2}\right), \quad \ldots\ldots (9)$$

falls wieder

$$\mathfrak{S} = c\,[\mathfrak{e}, \mathfrak{h}]$$

gesetzt wird. Integriert man über ein endliches Volumen, so erhält man ähnlich wie in §2

$$\frac{\partial}{\partial t}\left\{\int \frac{\mathfrak{e}\mathfrak{d} + \mathfrak{h}\mathfrak{b}}{2}\, d\tau\right\} = \int \mathfrak{S}_n\, d\sigma - \int \mathfrak{e}\mathfrak{i}\, d\tau.$$

Diese Gleichung spricht für den hier betrachteten Fall das Energieprinzip aus. Man sieht, dass der Vektor der Energieströmung lediglich von den Feldstärken abhängt, und zwar genau wie in dem Falle, dass polarisierbare Körper fehlen, und dass man $\frac{\mathfrak{e}\mathfrak{d}}{2}$ als die Dichte der elektrischen, $\frac{\mathfrak{h}\mathfrak{b}}{2}$ als die Dichte der magnetischen Energie anzusehen hat.

Wir können die Energiedichte des elektromagnetischen Feldes in folgende Anteile w, w_e und w_m zerlegen:

$$w = \frac{\mathfrak{e}^2 + \mathfrak{h}^2}{2}$$

$$w_e = \frac{1}{\varepsilon-1}\,\frac{\mathfrak{p}^2}{2}$$

$$w_m = \frac{1}{\mu-1}\,\frac{\mathfrak{m}^2}{2}.$$

Diese Zerlegung ist vom physikalischen Standpunkte aus mit Rücksicht auf die Lorentz'sche Theorie geboten. Es ist nämlich w die rein elektromagnetische Energiedichte, welche dem Feld auch dann zukäme, wenn polarisierbare Körper fehlten. w_e ist die Dichte der Energie, die man aufwenden muss, um dem Medium entgegen den elastischen Kräften zwischen gebundenen Elektrizitäten und Materie die Polarisation $\mathfrak{p}$ zu erteilen. Diese Energie haftet also der Materie an und hat mit dem elektromagnetischen Felde direkt nichts zu thun; sie braucht nicht als elektromagnetische Energie aufgefasst zu werden, sondern ist nur mit ihr vermöge der Eigenschaften der Materie verknüpft.[x] Das Gleiche gilt bezüglich w_m. Wir können für (9) schreiben:

$$\mathfrak{e}\mathfrak{i} = -\operatorname{div}\mathfrak{S} - \frac{\partial w}{\partial t} - \frac{\partial w_e}{\partial t} - \frac{\partial w_m}{\partial t} \quad \ldots (9a)$$

<u>Auf ruhende Körper ausgeübte ponderomotorische Kräfte.</u> Wir suchen diese Kräfte zunächst für den Spezialfall auf, dass nur ein elektrostatisches Feld vorhanden ist. Wir fragen nach der Arbeit A', welche das Feld bei einer unendlich kleinen Verrückung δ der materiellen Teilchen liefert. Aus dieser Arbeit folgen dann die auf die Materie wirkenden ponderomotorischen Kräfte. Aus dem Energieprinzip folgt

$$A' = -\delta\left\{\int \frac{\mathfrak{e}\mathfrak{d}}{2}\, d\tau\right\},$$

wobei „δ" die Änderung bedeutet, welche die hinter δ stehende Grösse infolge der

---

[x] Die eingeführte Zerlegung ist insbesondere gerechtfertigt mit Rücksicht auf den Fall, dass die Gleichungen (8) durch kompliziertere Beziehungen zu ersetzen sind. Es behält dann w seine Form bei, während die Ausdrücke für w_e und w_m durch kompliziertere zu ersetzen sind.

a result of the displacement. Because of the third of equations (Ia′), $\mathfrak{e}$ can be presented in the form −grad φ, so that, because of the first of equations (8) and the second of equations (Ia′), the energy of the field can be written in the form

$$\int \frac{1}{2} \varepsilon \operatorname{grad}^2 \varphi \, d\tau$$

as well as in the form

$$\int \frac{1}{2} \varphi \rho \, d\tau.$$

Thus,

$$E = \int (\varphi \rho - \frac{1}{2} \varepsilon \operatorname{grad}^2 \varphi) d\tau$$

also represents the energy of the system. The latter representation has a property that makes it much easier to construct the variation that is being sought. Namely, if one varies φ while the values for ρ and ε are kept constant, then the variation for the total system vanishes. For we have

$$\delta_\varphi E = \int \left[\rho \delta\varphi - \varepsilon \left(\frac{\partial \varphi}{\partial x} \delta \frac{\partial \varphi}{\partial x} + \cdot + \cdot \right) \right] d\tau$$

$$= \int \left[\rho \delta\varphi + \left(\mathfrak{d}_x \frac{\partial \delta\varphi}{\partial x} + \cdot + \cdot \right) \right] d\tau$$

$$= \int [\rho - \operatorname{div} \mathfrak{d}] \delta\varphi \, d\tau = 0$$

Thus, in order to find the variation of E that corresponds to an arbitrary virtual displacement, one need only vary ρ and ε, so that one has

$$\delta E = \int \left(\varphi \delta \rho - \frac{1}{2} \mathfrak{e}^2 \delta \varepsilon \right) d\tau.$$

Because of the indestructibility of the conduction electricity, and because the latter is displaced together with the matter in the static problem under consideration, we have

$$\delta \rho = -\operatorname{div} (\rho \mathfrak{s}),$$

where $\mathfrak{s}$ denotes the vector of the infinitely small spatial displacement of the system. We determine the variation of ε on the assumption that the dielectric constant ε of a material particle does not change during the displacement; by making this assumption, one excludes the phenomena of electrostriction from consideration. After the displacement one finds at the location of the radius vector $\mathfrak{r}$ the material particle that was at the location $\mathfrak{r} - \mathfrak{s}$ before the displacement. Therefore, given the indicated assumption,

$$\delta \varepsilon = -\mathfrak{s} \operatorname{grad} \varepsilon,$$

so that we have [21]

$$\delta E = \int \left(-\varphi \operatorname{grad} (\rho \mathfrak{s}) + \frac{1}{2} \mathfrak{e}^2 \operatorname{grad} \varepsilon \mathfrak{s} \right) d\tau.$$

To find the X-coordinate $\mathfrak{f}_x$ of the force acting on substance per unit volume, we have to consider only that part $\delta_x E$ of δE that corresponds to the component $\mathfrak{s}_x$ of the elementary displacement. After having transformed the first term of the integral by means of integration by parts, one gets [22]

$$\delta_x E = \int \left(-\mathfrak{e}_x \rho + \frac{1}{2} \mathfrak{e}^2 \frac{\partial \varepsilon}{\partial x} \right) \mathfrak{s}_x \, d\tau.$$

Verrückung erleidet. Wegen der dritten der Gleichungen (Ia') ist $\mathfrak{e}$ in der Form $-\mathrm{grad}\,\varphi$ darstellbar, sodass die Energie des Feldes wegen der ersten der Gleichungen (8) und der zweiten der Gleichungen (Ia') sowohl in der Form

$$~~\int \tfrac{1}{2}\varepsilon \tfrac{\partial\varphi}{\partial x}~~ \quad \int \frac{1}{2}\varepsilon\, \mathrm{grad}^2 \varphi \, d\tau$$

als auch in der Form

$$\int \frac{1}{2}\varphi\varrho \, d\tau$$

geschrieben werden kann. Es stellt also auch

$$\mathfrak{E} = \int \left(\varphi\varrho - \frac{1}{2}\varepsilon\, \mathrm{grad}^2 \varphi\right) d\tau$$

die Energie des Systems dar. Die letztere Darstellung hat eine Eigenschaft, welche die Bildung der gesuchten Variation sehr erleichtert. Wird nämlich φ bei festgehaltenen Werten von ϱ und ε variiert, so verschwindet die Variation für das Gesamtsystem. Es ist nämlich

$$\delta_\varphi \mathfrak{E} = \int \left[\varrho\,\delta\varphi - \varepsilon\left(\frac{\partial\varphi}{\partial x}\,\delta\frac{\partial\varphi}{\partial x} + \cdot + \cdot\right)\right] d\tau$$

$$= \int \left[\varrho\,\delta\varphi + \left(\mathfrak{d}_x \frac{\partial\,\delta\varphi}{\partial x} + \cdot + \cdot\right)\right] d\tau$$

$$= \int \left[\varrho - \mathrm{div}\,\mathfrak{d}\right] \delta\varphi \, d\tau = 0$$

Um also die einer beliebigen virtuellen Verschiebung entsprechende Variation von $\mathfrak{E}$ zu finden, braucht man nur ϱ und ε zu variieren, sodass man hat

$$\delta\mathfrak{E} = \int \left(\varphi\,\delta\varrho - \frac{1}{2}\mathfrak{e}^2\,\delta\varepsilon\right) d\tau$$

Wegen der Unzerstörbarkeit der Leitungselektrizität, und weil letztere bei dem behandelten statischen Problem mit der Materie zusammen verschoben wird, ist

$$\delta\varrho = -\,\mathrm{div}(\varrho\mathfrak{s}),$$

wobei mit $\mathfrak{s}$ der Vektor der unendlich kleinen räumlichen Verschiebung des Systems bezeichnet ist. Die Variation von ε bestimmen wir unter der Annahme, dass sich die Dielektrizitätskonstante ε eines materiellen Teilchens bei der Verrückung nicht ändere; durch diese Annahme werden die Erscheinungen der Elektro-Striktion von der Betrachtung ausgeschlossen. Es befindet sich nach der Verschiebung am Orte vom Radiusvektor $\mathfrak{r}$ das materielle Teilchen, welches vor der Verschiebung am Orte $\mathfrak{r} - \mathfrak{s}$ war. Deshalb ist unter der angegebenen Voraussetzung

$$\delta\varepsilon = -\,\mathfrak{s}\,\mathrm{grad}\,\varepsilon,$$

sodass man hat

$$\delta\mathfrak{E} = \int \left(-\varphi\,\mathrm{grad}(\varrho\mathfrak{s}) + \frac{1}{2}\mathfrak{e}^2\,\mathrm{grad}\,\varepsilon\,\mathfrak{s}\right) d\tau.$$

Um nun die X-Koordinate $\mathfrak{f}_x$ der pro Volumeneinheit auf die Substanz wirkenden Kraft zu finden, brauchen wir nur den Teil $\delta_x\mathfrak{E}$ von $\delta\mathfrak{E}$ zu betrachten, welcher der Komponente $\mathfrak{s}_x$ der elementaren Verrückung ~~parallel~~ entspricht. Es folgt, nachdem man den ersten Term des Integrals durch partielle Integration umgeformt hat

$$\delta_x\mathfrak{E} = \int \left(-\mathfrak{e}_x\varrho + \frac{1}{2}\mathfrak{e}^2\frac{\partial\varepsilon}{\partial x}\right)\mathfrak{s}_x \, d\tau$$

Da ~~man aber~~ $-\delta_x\mathfrak{E}$ die der Verschiebung $\mathfrak{s}_x$ entsprechende, auf die Materie

Since $-\delta_x E$ is the work transferred from field to matter corresponding to the displacement $\mathfrak{s}_x$, we have to set

$$\delta_x E = \int \mathfrak{f}_x \mathfrak{s}_x d\tau,$$

where $\mathfrak{f}_x$ denotes the force that the field exerts upon the matter in the direction of the X-coordinate. Accordingly,[23]

$$\mathfrak{f}_x = \mathfrak{e}_x \rho - \frac{1}{2}\mathfrak{e}^2 \frac{\partial \varepsilon}{\partial x}.$$

Here, too, $\mathfrak{f}_x$ can be represented in a form that makes it evident that the momentum law is satisfied. For

$$\frac{1}{2}\mathfrak{e}^2 \frac{\partial \varepsilon}{\partial x} = \frac{\partial}{\partial x}\left(\frac{1}{2}\varepsilon \mathfrak{e}^2\right) - \varepsilon\left(\mathfrak{e}_x \frac{\partial \mathfrak{e}_x}{\partial x} + \mathfrak{e}_y \frac{\partial \mathfrak{e}_y}{\partial x} + \mathfrak{e}_z \frac{\partial \mathfrak{e}_z}{\partial x}\right),$$

or, according to the second and the third of equations (Ia′) and the first of equations (8),

$$\frac{1}{2}\mathfrak{e}^2 \frac{\partial \varepsilon}{\partial x} = \frac{\partial}{\partial x}\left(\frac{\mathfrak{e}\mathfrak{d}}{2} - \mathfrak{e}_x \mathfrak{d}_x\right) - \frac{\partial}{\partial y}(\mathfrak{e}_x \mathfrak{d}_y) - \frac{\partial}{\partial z}(\mathfrak{e}_x \mathfrak{d}_z) + \mathfrak{e}_x \rho.$$

Therefore, if one sets

$$p'_{xx} = \frac{\mathfrak{e}\mathfrak{d}}{2} - \mathfrak{e}_x \mathfrak{d}_x \quad p'_{xy} = -\mathfrak{e}_x \mathfrak{d}_y \quad p'_{xz} = -\mathfrak{e}_x \mathfrak{d}_z \text{ etc.,}$$

one obtains

$$\mathfrak{f}_x = -\frac{\partial p'_{xx}}{\partial x} - \frac{\partial p'_{xy}}{\partial y} - \frac{\partial p'_{xz}}{\partial z}.$$

Analogous expressions are obtained for $\mathfrak{f}_y$ and $\mathfrak{f}_z$. The 9 quantities p'_{xx} etc. are the "Maxwell pressure forces" for the special case of the electrostatic field. Totally analogous expressions are obtained for the case of the magnetostatic problem insofar as the second of equations (8) correctly describes the relation between $\mathfrak{h}$ and $\mathfrak{b}$. Thus, one obtains the ponderomotive forces for static problems in general if one sets

$$\left.\begin{array}{ll} p'_{xx} = \dfrac{\mathfrak{e}\mathfrak{d} + \mathfrak{h}\mathfrak{b}}{2} - \mathfrak{e}_x \mathfrak{d}_x - \mathfrak{h}_x \mathfrak{b}_x & p'_{xy} = -\mathfrak{e}_x \mathfrak{d}_y - \mathfrak{h}_x \mathfrak{b}_y \\ p'_{xz} = -\mathfrak{e}_x \mathfrak{d}_z - \mathfrak{h}_x \mathfrak{b}_z \text{ etc.} & \end{array}\right\} \quad (10)$$

As will be shown in what follows, from H. A. Lorentz's standpoint it can easily be recognized that these expressions are also valid for electrodynamic processes in bodies at rest. In that case, we have to bear in mind that the pressure forces given here consist of components of physically diverse kinds that can easily be separated. That is to say, we have, first, those pressure forces by means of which we represented in §2 the ponderomotive effects of the electromagnetic field ($\mathfrak{h}$, $\mathfrak{e}$) on the set of all of the electrical densities of matter. But, second, electrical polarization brings about pressure forces due to the circumstance that the (infinitesimally small) elastically displaced electrical continua of polarization exert forces, whose nature we cannot, in general, characterize more precisely, on the material particles connected with them. Obviously, these forces depend only on the polarization and the dielectric constant but not on the field strength. From the physical standpoint it is natural, therefore, to decompose the Maxwell pressure forces into components in the following way

vom Felde übertragene Arbeit ist, so haben wir

$$-\delta_x E = \int f_x \delta_x \, d\tau$$

zu setzen, wobei f_x die in Richtung der X-Koordinate vom Felde auf die Materie ausgeübte Kraft bedeutet. Es ist demnach

$$f_x = \mathfrak{e}_x \varrho - \frac{1}{2} \mathfrak{e}^2 \frac{\partial \varepsilon}{\partial x}.$$

Auch hier lässt sich f_x in einer Form darstellen, welche erkennen lässt, dass dem Impulssatz Genüge geleistet ist. Es ist nämlich

$$\frac{1}{2} \mathfrak{e}^2 \frac{\partial \varepsilon}{\partial x} = \frac{\partial}{\partial x}\left(\frac{1}{2} \varepsilon \mathfrak{e}^2\right) - \varepsilon\left(\mathfrak{e}_x \frac{\partial \mathfrak{e}_x}{\partial x} + \mathfrak{e}_y \frac{\partial \mathfrak{e}_y}{\partial x} + \mathfrak{e}_z \frac{\partial \mathfrak{e}_z}{\partial x}\right),$$

oder nach der zweiten und dritten der Gleichungen (Ia') und der ersten der Gleichungen (8)

$$\frac{1}{2} \mathfrak{e}^2 \frac{\partial \varepsilon}{\partial x} = \frac{\partial}{\partial x}\left(\frac{\mathfrak{e}\mathfrak{d}}{2} - \mathfrak{e}_x \mathfrak{d}_x\right) - \frac{\partial}{\partial y}(\mathfrak{e}_x \mathfrak{d}_y) - \frac{\partial}{\partial z}(\mathfrak{e}_x \mathfrak{d}_z) + \mathfrak{e}_x \varrho.$$

Setzt man daher

$$p'_{xx} = \frac{\mathfrak{e}\mathfrak{d}}{2} - \mathfrak{e}_x \mathfrak{d}_x \qquad p'_{xy} = -\mathfrak{e}_x \mathfrak{d}_y \qquad p'_{xz} = -\mathfrak{e}_x \mathfrak{d}_z \text{ etc.},$$

so erhält man

$$f_x = -\frac{\partial p'_{xx}}{\partial x} - \frac{\partial p'_{xy}}{\partial y} - \frac{\partial p'_{xz}}{\partial z}.$$

Analoge Ausdrücke erhält man ~~erst~~ für f_y und f_z. Die 9 Grössen p'_{xx} etc. sind die „Maxwell'schen Druck-Kräfte" für den Spezialfall des elektrostatischen Feldes. Ganz entsprechende Ausdrücke bekommt man für den Fall des magnetostatischen Problems, soweit die ~~Beziehung zu~~ zweite der Gleichungen (8) die Beziehung zwischen $\mathfrak{h}$ und $\mathfrak{b}$ richtig wiedergibt. Man erhält also die ponderomotorischen Kräfte für statische Probleme überhaupt, indem man setzt

$$p'_{xx} = \frac{\mathfrak{e}\mathfrak{d} + \mathfrak{h}\mathfrak{b}}{2} - \mathfrak{e}_x \mathfrak{d}_x - \mathfrak{h}_x \mathfrak{b}_x \qquad p'_{xy} = -\mathfrak{e}_x \mathfrak{d}_y - \mathfrak{h}_x \mathfrak{b}_y \qquad p'_{xz} = -\mathfrak{e}_x \mathfrak{d}_z - \mathfrak{h}_x \mathfrak{b}_z \text{ etc.} \quad (10)$$

Es ist vom ~~Standpunkt der~~ H. A. Lorentz'schen ~~Auffassung aus~~ Standpunkt aus, wie im Nachfolgenden gezeigt wird, leicht einzusehen, dass diese Ausdrücke auch für elektrodynamische Vorgänge in ruhenden Körpern ihre Gültigkeit behalten. Wir haben da zu beachten, dass die hier angegebenen Druckkräfte aus physikalisch verschiedenartigen Bestandteilen bestehen, die sich leicht trennen lassen. Es bestehen nämlich erstens ~~des elektromagnetischen Feldes~~ diejenigen Druckkräfte, mittelst welcher wir im §2 die ponderomotorischen Wirkungen des elektromagnetischen Feldes ($\mathfrak{h}$, $\mathfrak{e}$) auf die Gesamtheit der elektrischen Dichten der Materie dargestellt haben. Zweitens aber bringt die elektrische Polarisation dadurch Druckkräfte zustande, dass die (unendlich wenig) elastisch verschobenen elektrischen Kontinua der Polarisation auf die mit ihnen verbundenen materiellen Teilchen Kräfte ausüben, deren Natur wir allgemein nicht näher charakterisieren können. Diese Kräfte hängen offenbar nur von der Polarisation und der Dielektrizitätskonstante aber nicht von der Feldstärke ab. Analog verhält es sich mit der magnetischen Polarisation. Es ist daher vom physikalischen Standpunkte aus natürlich, die Maxwell'schen Druckkräfte in folgender Weise in Komponenten zu zerlegen

$$p'_{xx} = p_{xx} + p^{(e)}_{xx} + p^{(m)}_{xx} \qquad p'_{xy} = p_{xy} + p^{(e)}_{xy} + p^{(m)}_{xy} \text{ etc.,}$$

where

$$\left.\begin{aligned} p_{xx} &= \frac{\mathfrak{e}^2 + \mathfrak{h}^2}{2} - \mathfrak{e}_x^2 - \mathfrak{h}_x^2 \qquad p_{xy} = -\mathfrak{e}_x\mathfrak{e}_y - \mathfrak{h}_x\mathfrak{h}_y \text{ etc.} \\ p^{(e)}_{xx} &= \frac{1}{\varepsilon - 1}\left(\frac{\mathfrak{p}^2}{2} - \mathfrak{p}_x^2\right) \qquad p^{(e)}_{xy} = \frac{1}{\varepsilon - 1}\mathfrak{p}_x\mathfrak{p}_y \text{ etc.} \\ p^{(m)}_{xx} &= \frac{1}{\mu - 1}\left(\frac{\mathfrak{m}^2}{2} - \mathfrak{m}_x^2\right) \qquad p^{(m)}_{xy} = \frac{1}{\mu - 1}\mathfrak{m}_x\mathfrak{m}_y \text{ etc.} \end{aligned}\right\} \quad (10a)$$

As is shown in §2, the first system (p_{xx} etc.) of pressure forces is valid for arbitrary dynamic problems. For the second and the third system ($p^{(e)}_{xx}$ etc. and $p^{(m)}_{xx}$ etc.), this general validity follows from the circumstance that the elastic forces of polarization do not depend on whether the deformations that make up the polarization are temporally invariant or not. From this it follows that the Maxwell pressure forces retain their meaning in the case of dynamical problems as well.

As a glance at equations (7) shows, in order to obtain the forces exerted by the electromagnetic field on the matter in bodies at rest, all that we need know, in addition, is the momentum of the electromagnetic field. In order for the momentum conservation law not to be violated by electrodynamics, those forces must be determined by equations of the form of equations (7). But just as in nonpolarizable media, the momentum of the field in polarized media will be determined by the field strengths $\mathfrak{e}$ and $\mathfrak{h}$ alone. For it is impossible to see why momentum and energy flux should correspond to the elastic forces of polarization. Therefore, if one again sets [24] $\mathfrak{s}_x = c[\mathfrak{e}, \mathfrak{h}]$, then—at least within the range of validity of equations (8)—one has to set

$$\mathfrak{f}_x = -\frac{\partial p'_{xx}}{\partial x} - \frac{\partial p'_{xy}}{\partial y} - \frac{\partial p'_{xz}}{\partial z} - \frac{1}{c^2}\frac{\partial \mathfrak{s}_x}{\partial t} \text{ etc.} \qquad \ldots (11)$$

If one inserts the expressions for p'_{xx} etc. from (10) or (10a) into this formula, and transforms by means of equations (Ia), one therefore obtains for the ponderomotive force the physically intuitive formula*

$$\left.\begin{aligned} \mathfrak{f} = \mathfrak{e}\rho + \frac{1}{c}[\mathfrak{i}, \mathfrak{h}] - \frac{1}{2}\text{grad}(\mathfrak{e}\,\mathfrak{p}) + (\mathfrak{p}\nabla)\,\mathfrak{e} + \frac{1}{c}[\dot{\mathfrak{p}}, \mathfrak{h}] \\ - \frac{1}{2}\text{grad}(\mathfrak{h}, \mathfrak{m}) + (\mathfrak{m}\nabla)\,\mathfrak{h} - \frac{1}{c}[\dot{\mathfrak{m}}, \mathfrak{e}] \end{aligned}\right\} \quad (11a)$$

In this formula only the terms $-\frac{1}{2}$ grad ($\mathfrak{e}\mathfrak{p}$) and $-$grad($\mathfrak{h}\mathfrak{m}$) were physically opaque. They owe their appearance to the assumption, which we introduced above, that during an infinitesimally small displacement of the matter the dielectric constant of the material particle always remains unchanged, even in the case where the particle

---

* $(\mathfrak{p}\nabla)\mathfrak{e}$ is the vector whose x-component is $\mathfrak{p}_x\frac{\partial \mathfrak{e}_x}{\partial x} + \mathfrak{p}_y\frac{\partial \mathfrak{e}_x}{\partial y} + \mathfrak{p}_z\frac{\partial \mathfrak{e}_x}{\partial z}$. The physical cause of the appearance of this term is that the polarization electricities of a material particle are situated at somewhat different locations, so that the field strengths acting on them are somewhat different.

$$\left.\begin{aligned}
&p'_{xx} = p_{xx} + p^{(e)}_{xx} + p^{(m)}_{xx} \qquad p'_{xy} = p_{xy} + p^{(e)}_{xy} + p^{(m)}_{xy} \quad \text{etc.,}\\
&\text{wobei gesetzt ist}\\
&p_{xx} = \frac{\mathfrak{e}^2 + \mathfrak{h}^2}{2} - \mathfrak{e}_x^2 - \mathfrak{h}_x^2 \qquad p_{xy} = -\mathfrak{e}_x\mathfrak{e}_y - \mathfrak{h}_x\mathfrak{h}_y \quad \text{etc.}\\
&p^{(e)}_{xx} = \frac{1}{\varepsilon-1}\left(\frac{\mathfrak{J}^2}{2} - \mathfrak{J}_x^2\right) \qquad p^{(e)}_{xy} = -\frac{1}{\varepsilon-1}\mathfrak{J}_x\mathfrak{J}_y \quad \text{etc.}\\
&p^{(m)}_{xx} = \frac{1}{\mu-1}\left(\frac{\mathfrak{M}^2}{2} - \mathfrak{M}_x^2\right) \qquad p^{(m)}_{xy} = -\frac{1}{\mu-1}\mathfrak{M}_x\mathfrak{M}_y \quad \text{etc.}
\end{aligned}\right\} \quad (10a)$$

Das erste System (p_{xx} etc.) von Druckkräften hat, wie in §2 gezeigt ist, für beliebige dynamische Probleme Gültigkeit. Von dem zweiten und dritten Systeme ($p^{(e)}_{xx}$ etc. und $p^{(m)}_{xx}$ etc.) folgt diese allgemeine Gültigkeit daraus, dass die elastischen Kräfte der Polarisation unabhängig davon sind, ob die Deformationen, welche die Polarisation ausmachen, zeitlich unveränderlich sind oder nicht. Es folgt hieraus, dass die Maxwellschen Druckkräfte auch im Falle dynamischer Probleme ihre Bedeutung behalten.

Um die in ruhenden Körpern ~~wirkenden~~ vom elektromagnetischen Felde ~~ausgeü.~~ auf die Materie ausgeübten Kräfte zu erhalten, müssen wir ~~nur noch~~, wie ein Blick auf die Gleichungen (7) lehrt, nur noch den Impuls des elektromagnetischen Feldes kennen. Denn es müssen, damit der Satz von der Erhaltung der Bewegungsgrösse durch die Elektrodynamik nicht verletzt werde, jene Kräfte durch Gleichungen von der Form der Gleichungen (7) bestimmt sein. Der Impuls des Feldes wird aber in polarisierten Medien genau so durch die Feldstärken $\mathfrak{e}$ und $\mathfrak{h}$ allein bestimmt sein, wie in nicht polarisierbaren Medien. Denn es ist nicht einzusehen, wieso den elastischen Kräften der Polarisation Bewegungsgrösse und Energieströmung entsprechen ~~zukommen~~ sollte. Setzt man daher wieder $\mathfrak{S}_x = c[\mathfrak{e}, \mathfrak{h}]$, so ist – wenigstens innerhalb des Gültigkeitsbereiches der Gleichungen (8) zu setzen

$$\mathfrak{f}_x = -\frac{\partial p'_{xx}}{\partial x} - \frac{\partial p'_{xy}}{\partial y} - \frac{\partial p'_{xz}}{\partial z} - \frac{1}{c^2}\frac{\partial \mathfrak{S}_x}{\partial t} \quad \text{etc.} \qquad \ldots (11)$$

Setzt man in diese Formel die Ausdrücke für p'_{xx} etc. aus (10) oder (10a) ein, und formt man mittelst der Gleichungen (Ia) um, so erhält man für die ponderomotorische Kraft die physikalisch anschauliche Formel*

$$\left.\begin{aligned}
\mathfrak{f} = \mathfrak{e}\rho + \frac{1}{c}[\mathfrak{i}, \mathfrak{h}] - \frac{1}{2}\operatorname{grad}(\mathfrak{e}\mathfrak{J}) + (\mathfrak{J}\nabla)\mathfrak{e} + \frac{1}{c}[\dot{\mathfrak{J}}, \mathfrak{h}] \; \text{~~}{+\frac{1}{c}[\mathfrak{i}, \mathfrak{h}]}\text{~~}\\
- \frac{1}{2}\operatorname{grad}(\mathfrak{h}\mathfrak{M}) + (\mathfrak{M}\nabla)\mathfrak{h} - \frac{1}{c}[\dot{\mathfrak{M}}, \mathfrak{e}]
\end{aligned}\right\} \quad (11a)$$

In dieser Formel entbehrten nur die Glieder $-\frac{1}{2}\operatorname{grad}(\mathfrak{e}\mathfrak{J})$ und $-\frac{1}{2}\operatorname{grad}(\mathfrak{h}\mathfrak{M})$ der physikalischen Durchsichtigkeit. Sie verdanken ihr Auftreten der oben eingeführten Voraussetzung, dass bei einer unendlich kleinen Verrückung der Materie die Dielektrizitätskonstante des materiellen Teilchens stets ungeändert bleibe, auch dann, wenn das Teilchen bei der Verrückung sein Volumen ändert. Diese Glieder ~~sind aber auch von~~

---

* $(\mathfrak{J}\nabla)\mathfrak{e}$ ist der Vektor, dessen x-Komponente $\mathfrak{J}_x\frac{\partial \mathfrak{e}_x}{\partial x} + \mathfrak{J}_y\frac{\partial \mathfrak{e}_x}{\partial y} + \mathfrak{J}_z\frac{\partial \mathfrak{e}_x}{\partial z}$ ist. Die physikalische Ursache des Auftretens dieses Gliedes ist die, dass die Polarisationselektrizitäten eines materiellen Teilchens sich an etwas verschiedenen Stellen befinden, sodass die auf sie wirkenden Feldstärken etwas verschieden sind.

changes its volume in the course of the displacement. Thus, these terms correspond to a physically unjustified assumption. But they are also of subordinate interest because they are not capable of producing motion in incompressible substances, but only changes in pressure.

§4. Lorentz's Equations for (Slowly) Moving Media

As in the case of bodies at rest, we again start out from the fundamental equations (I) in §1, which, however, for reasons analyzed in §2, we must modify by assuming the presence of several continua of electrical density, which are to be viewed partly as carriers of polarization and partly as carriers of the electrical conduction currents and corresponding charges. In addition, the third and the fourth of equations (I) are to be supplemented by adding terms corresponding to the magnetic polarization currents. But because of the duality of electric and magnetic processes,[25] it suffices to set up the first two equations for our case. They read

$$\text{curl } \mathfrak{h} = \frac{1}{c}\left(\dot{\mathfrak{e}} + \sum (\mathfrak{q}_g \rho_g + \sum \mathfrak{q}_l \rho_l\right)$$

$$\text{div } \mathfrak{e} = \sum \rho_g + \sum \rho_l$$

If, for the sake of brevity, we denote the sum of the second and the third term within the bracket on the left-hand side by $\mathfrak{a}$, then $\mathfrak{a}$ is the vector of the total electrical current with respect to an observer at rest. Our task is to express $\mathfrak{a}$ by means of the vectors $\mathfrak{q}$, $\mathfrak{i}$ (conduction current vector) and $\mathfrak{p}$ (polarization vector), the last two of which can be sharply defined only for bodies at rest, that is, in our case, for an observer moving with the material element under consideration.

If one chooses an infinitesimally small, plane surface element σ *at rest,* with a normal n, then, according to the definition of $\mathfrak{a}$, the product $\mathfrak{a}_n \sigma$ is equal to the quantity of electricity traversing σ per unit time. If we succeed in finding this for an arbitrarily oriented surface element, then we also have along with it the expression for $\mathfrak{a}$. Now we inquire about the quantity of electricity that, per unit time, traverses that particular surface element *moving with the matter* that coincides with σ at the beginning of the unit of time under consideration, but takes up the position σ' at the time $t + dt$. The total amount of conduction electricity passing through this element is

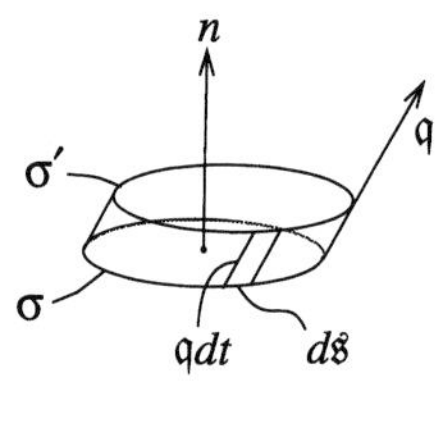

$$\mathfrak{i}_n \sigma dt.$$

But the total amount of polarization electricity passing through the comoving element is equal to the change that the expression $\mathfrak{p}_n \sigma$ undergoes during the unit time considered, and thus equal to $\mathfrak{p}_n{}' \sigma' - \mathfrak{p}_n \sigma$. This change <according to the calculation rule derived in the appendix ( )> is equal to [26] *

$$(\dot{\mathfrak{p}} + \mathfrak{q} \text{ div } \mathfrak{p} - \text{curl } [\mathfrak{q}\mathfrak{p}])_n \, \sigma dt,$$

where the index at the bracket means that one is to take the component along the

* One finds this difference by extending div $\mathfrak{p}$ over the volume described by the surface element during time *dt*; for we have, on the one hand,

$$\int \text{div } \mathfrak{p} d\tau = \mathfrak{q}_n \sigma dt \cdot \text{div } \mathfrak{p},$$

and, on the other hand, $\int \text{div } \mathfrak{p} d\tau = \int \mathfrak{p}_n d\sigma = \mathfrak{p}'_n \sigma' - \mathfrak{p}_n \sigma + dt \int [d\mathfrak{s}, \mathfrak{q}] \, \mathfrak{p}$,

where for the latter line intergral one can set $-\int [\mathfrak{q}, \mathfrak{p}] \, d\mathfrak{s} = \text{rot } [\mathfrak{q}, \mathfrak{p}] \, \sigma$ (according to Stoke's theorem.)

entsprechen also einer physikalisch nicht begründeten Annahme. Sie besitzen aber auch ein untergeordnetes Interesse, da sie in inkompressibeln Stoffen nicht Bewegungen sondern nur Druckänderungen der Materie herbeizuführen vermögen.

§4. Lorentz'sche Gleichungen für (langsam) bewegte Medien.

Wie im Falle ruhender Körper gehen wir wieder von den Grundgleichungen (I) des §1 aus, an denen wir jedoch aus den in §2 auseinandergesetzten Gründen die Aenderungen anzubringen haben, dass wir mehrere Kontinuen elektrischer Dichte als vorhanden annehmen, welche teils als Träger der Polarisation, teils als Träger der elektrischen Leitungsströme und zugehörigen Ladungen aufzufassen sind. Die dritte und vierte der Gleichungen (I) sind ausserdem um Glieder zu ergänzen, welche den magnetischen Polarisationsströmen entsprechen. Es genügt aber wegen der Dualität der elektrischen und magnetischen Vorgänge die ersten beiden Gleichungen für unseren Fall aufzustellen. Diese lauten

$$\mathrm{curl}\,\mathfrak{h} = \frac{1}{c}\left(\mathfrak{i} + \sum \mathfrak{v}_g \varrho_g + \sum \mathfrak{v}_e \varrho_e\right)$$

$$\mathrm{div}\,\mathfrak{e} = \sum \varrho_g + \sum \varrho_e$$

Bezeichnen wir zur Abkürzung die Summe des zweiten und dritten Gliedes der Klammer der rechten Seite mit $\mathfrak{s}$, so ist der Vektor des elektrischen Gesamtstromes inbezug auf einen ruhenden Beobachter; es ist unsere Aufgabe $\mathfrak{s}$ durch die Vektoren $\mathfrak{v}$, $\mathfrak{i}$ (Leitungsstromvektor) und $\mathfrak{g}$ (Polarisationsvektor) auszudrücken, von denen die beiden letzten lediglich für ruhende Körper, das heisst in unserem Falle für einen mit dem betrachteten materiellen Element bewegten Beobachter scharf definiert werden können.

Wählt man ein ruhendes, unendlich kleines, ebenes Flächenelement σ mit Normale n, so ist nach der Definition von $\mathfrak{s}$ das Produkt $\mathfrak{s}_n \sigma$ gleich der durch σ pro Zeiteinheit hindurchtretenden Elektrizitätsmenge. Gelingt es, diese für ein beliebig orientiertes Flächenelement zu finden, so besitzen wir damit auch den Ausdruck für $\mathfrak{s}$. Wir fragen nun zunächst nach der durch dasjenige mit der Materie bewegte Flächenelement pro Zeiteinheit hindurchtretenden Elektrizitätsmenge, welches zu Anfang der ins Auge gefassten Zeiteinheit mit σ zusammenfällt, in der Zeit $t + dt$ aber die Lage σ' annimmt. Die Summe der dieses Element passierenden Leitungselektrizität ist

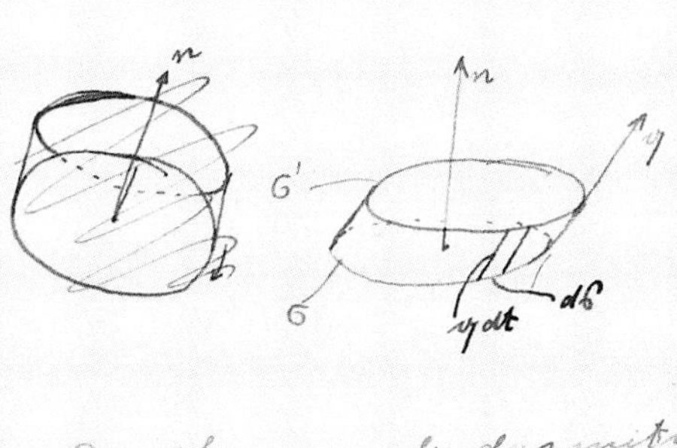

$$\mathfrak{i}_n \sigma\, dt$$

Die Summe der das mitbewegte Element passierenden Polarisationselektrizität ist aber gleich der Aenderung, welche der Ausdruck $\mathfrak{g}_n \sigma$ in der betrachteten Zeiteinheit erfährt, also gleich $\mathfrak{g}_n'\sigma' - \mathfrak{g}_n\sigma$. Diese Aenderung ist* gleich

$$\left(\dot{\mathfrak{g}} + \mathfrak{v}\,\mathrm{div}\,\mathfrak{g} - \mathrm{curl}[\mathfrak{v}\,\mathfrak{g}]\right)_n \sigma\, dt,$$

wobei der Index bei der Klammer bedeutet, dass die nach der Normale von σ genommene Komponente des in der Klammer angegebenen Vektors zu nehmen ist. Die Summe dieser beiden Glieder ist gleich der elektrischen

---

* Man findet diese Differenz, indem man $\mathrm{div}\,\mathfrak{g}$ über das von dem Flächenelement während dt beschriebene Volumen erstreckt; denn es ist einerseits

$$\int \mathrm{div}\,\mathfrak{g}\, d\tau = \mathfrak{v}_n \sigma\, dt \cdot \mathrm{div}\,\mathfrak{g},$$

andererseits

$$\int \mathrm{div}\,\mathfrak{g}\, d\tau = \int \mathfrak{g}_n\, d\sigma = \mathfrak{g}_n'\sigma' - \mathfrak{g}_n\sigma + dt\int [d\mathfrak{s}, \mathfrak{v}]\,\mathfrak{g},$$

wobei für letzteres Linien-Integral $\int [\mathfrak{v}, \mathfrak{g}]\, d\mathfrak{s} = \mathrm{rot}[\mathfrak{v}, \mathfrak{g}]\,\sigma$ gesetzt werden kann (nach dem Satz von Stokes).

(Continuation of the note)

Thus, one would have

$$\mathfrak{p}'_n\sigma' - \mathfrak{p}_n\sigma = \{\mathfrak{q}_n \operatorname{div} \mathfrak{p} - \operatorname{curl}_n [\mathfrak{q}, \mathfrak{p}]\}\, \sigma dt$$

if $\mathfrak{p}$ would vary only with location but not with time. In case of the latter circumstance, $\mathfrak{p}_n\sigma$ also experiences the change

$$\dot{\mathfrak{p}}_n \sigma dt$$

in the time element *dt*. The summing of these two expressions yields the expression given in the text.

(Fortsetzung zur Anmerkung) 13a

Man hätte also

$$\mathfrak{j}_n'\sigma' - \mathfrak{j}_n\sigma = \left\{ \mathfrak{v}_n \operatorname{div} \mathfrak{j} - \operatorname{curl}_n [\mathfrak{v}, \mathfrak{j}] \right\} \sigma dt$$

wenn $\mathfrak{j}$ nur örtlich, nicht aber zeitlich variierte. Der letztere Umstand bringt es mit sich, dass $\mathfrak{j}_n\sigma$ im Zeitelement dt ~~die~~ ausserdem die Änderung $\dot{\mathfrak{j}}_n \sigma dt$ erfährt. ~~Aus~~ Durch Summierung dieser ~~beiden~~ Ausdrücke entsteht der im Text angegebene Ausdruck.

normal of σ of the vector given in the bracket. The sum of these two terms is equal to the total electrical current flowing through the comoving surface element. The total current through the element *at rest* σ exceeds the former by the total quantity of electricity present in the space $\mathfrak{q}_n\sigma$ that is swept by the moving element σ in unit time. It is given by

$$\mathfrak{q}_n \operatorname{div} e\sigma.$$

Thus, the total current passing through the element at rest σ is equal to

$$\mathfrak{i} + \dot{\mathfrak{p}} + \mathfrak{q} \operatorname{div} \mathfrak{p} - \operatorname{curl} [\mathfrak{q},\mathfrak{p}] + \mathfrak{q} \operatorname{div} \mathfrak{e})_n\sigma$$

or, since div $\mathfrak{e}$ + div $\mathfrak{p}$ = div $\mathfrak{d}$ = ρ, it is equal to

$$(\mathfrak{i} + \dot{\mathfrak{p}} - \operatorname{curl} [\mathfrak{q}\mathfrak{p}] + \mathfrak{q}\rho)_n\sigma.$$

Thus, the expression within the bracket is equal to the vector of the total current that was previously designated, for the sake of brevity, by $\mathfrak{a}$. Thus in our case, the first of equations (I) assumes the form

$$\operatorname{curl} \mathfrak{h} = \frac{1}{c}\,(\dot{\mathfrak{e}} + \dot{\mathfrak{p}} - \operatorname{curl} [\mathfrak{q},\mathfrak{p}] + \mathfrak{i} + \mathfrak{q}\rho).$$

For the second of equations (I) one obtains again

$$\operatorname{div} \mathfrak{e} = -\operatorname{div} \mathfrak{p} + \rho,$$

but here, of course, ρ does not denote the total density of electricity as in (I) but *only* the density of the conduction electricity (also called "true electricity").

The third and the fourth equations read analogously, in accordance with the principle of duality. One has only to bear in mind that there is no magnetic analog to the electrical conduction current. Accordingly, the field equations read

$$\left.\begin{array}{l|l} c \operatorname{curl} \mathfrak{h} - \dot{\mathfrak{e}} = (\dot{\mathfrak{p}} - \operatorname{curl} [\mathfrak{q},\mathfrak{p}]) + (\mathfrak{i} + \mathfrak{q}\rho) & c \operatorname{curl} \mathfrak{e} + \dot{\mathfrak{h}} = -(\dot{\mathfrak{m}} - \operatorname{curl} [\mathfrak{q},\mathfrak{m}]) \\ \operatorname{div} \mathfrak{e} = -\operatorname{div} \mathfrak{p} + \rho & \operatorname{div} \mathfrak{h} = -\operatorname{div} \mathfrak{m} \end{array}\right\} \quad \text{(Ib)}$$

The form of these equations tempts one to introduce the vectors $\mathfrak{h} + \frac{1}{c}[\mathfrak{q},\mathfrak{p}]$, $\mathfrak{e} + \mathfrak{p}$, $\mathfrak{e} - \frac{1}{c}[\mathfrak{q}\mathfrak{m}]$, $\mathfrak{h} + \mathfrak{m}$ as field vectors and, in this way, to simplify the equations considerably. In this way one obtains Minkowski's form of the field equations, which agrees with the form of the field equations for bodies at rest, up to the term $\mathfrak{q}\rho$.[27] But we do not do this, for the vectors that are to be introduced in this way possess no simple physical meaning. What one gains in the simplicity of the fundamental equations through their introduction, one loses again through the fact that the equations that give the connection between the states of the bodies and the field vectors become more complicated. Taking into account that the force acting on a moving unit of electricity according to §2 is equal to $\mathfrak{e} + \frac{1}{c}[\mathfrak{q},\mathfrak{h}]$ ("electromotive force") and that the force acting on the moving magnetic charge unit* is equal to $\mathfrak{h} - \frac{1}{c}[\mathfrak{q}\mathfrak{e}]$ ("magnetomotive force"),[28] one obtains, analogously to equations (8′),

$$\left.\begin{aligned} \mathfrak{p} &= (\varepsilon - 1)\,(\mathfrak{e} + \tfrac{1}{c}[\mathfrak{q},\mathfrak{h}]) \\ \mathfrak{m} &= (\mu - 1)\,(\mathfrak{h} - \tfrac{1}{c}[\mathfrak{q},\mathfrak{e}]) \\ \mathfrak{i} &= \lambda(\mathfrak{e} + \tfrac{1}{c}[\mathfrak{q},\mathfrak{h}]) \end{aligned}\right\} \quad (12)$$

* One can prove this easily if one expands the considerations presented in §2 by introducing also a magnetic density and a magnetic current.

Gesamtströmung durch das mitbewegte ~~Element~~ Flächenelement. Die Gesamtströmung durch das <u>ruhende</u> Element σ übertrifft jene um diejenige Gesamt-Elektrizitätsmenge, welche sich in dem in der Zeiteinheit von dem bewegten Element σ bestrichenen Raume $\mathfrak{q}_n \sigma$ befindet. Sie ist gegeben durch

$$\mathfrak{q}_n \operatorname{div} \mathfrak{e}\, \sigma$$

Die Gesamtströmung durch das ruhende Element σ ist also gleich

$$\left(\mathfrak{i} + \dot{\mathfrak{p}} + \mathfrak{q} \operatorname{div} \mathfrak{p} - \operatorname{curl}[\mathfrak{q}, \mathfrak{p}] + \mathfrak{q} \operatorname{div} \mathfrak{e}\right)_n \sigma$$

oder, ~~gleich~~ da $\operatorname{div} \mathfrak{e} + \operatorname{div} \mathfrak{p} = \operatorname{div} \mathfrak{d} = \rho$ ist, gleich

$$\left(\mathfrak{i} + \dot{\mathfrak{p}} - \operatorname{curl}[\mathfrak{q}\,\mathfrak{p}] + \mathfrak{q}\rho\right)_n \sigma$$

Die Klammer ist also gleich demjenigen Vektor der Gesamtströmung, der vorhin kurz mit $\mathfrak{s}$ bezeichnet wurde. Die erste der Gleichungen (I) nimmt also in unserem Falle die Gestalt an

$$\operatorname{curl} \mathfrak{h} = \frac{1}{c}\left(\dot{\mathfrak{e}} + \dot{\mathfrak{p}} - \operatorname{curl}[\mathfrak{q}, \mathfrak{p}] + \mathfrak{i} + \mathfrak{q}\rho\right).$$

Für die zweite der Gleichungen (I) erhält man wieder

$$\operatorname{div} \mathfrak{e} = -\operatorname{div} \mathfrak{p} + \rho,$$

wobei aber natürlich hier ρ nicht wie in (I) die Gesamt-Dichte der Elektrizität, sondern <u>nur</u> die Dichte der Leitungselektrizität (auch „wahre Elektrizität" genannt) bezeichnet.

Die dritte und vierte Gleichung lautet analog gemäss dem Prinzip der Dualität. Man hat nur zu beachten, dass das magnetische Analogon zum elektrischen Leitungsstrom fehlt. Die Feldgleichungen lauten demnach

$$\left.\begin{array}{l|l} c \operatorname{curl} \mathfrak{h} - \dot{\mathfrak{e}} = (\dot{\mathfrak{p}} - \operatorname{curl}[\mathfrak{q}, \mathfrak{p}]) + (\mathfrak{i} + \mathfrak{q}\rho) & c \operatorname{curl} \mathfrak{e} + \dot{\mathfrak{h}} = -(\dot{\mathfrak{m}} - \operatorname{curl}[\mathfrak{q}, \mathfrak{m}]) \\ \operatorname{div} \mathfrak{e} = -\operatorname{div} \mathfrak{p} + \rho & \operatorname{div} \mathfrak{h} = -\operatorname{div} \mathfrak{m} \end{array}\right\} \text{(Ib)}$$

Die Form dieser Gleichungen ladet dazu ein, die Vektoren $\mathfrak{h} + \frac{1}{c}[\mathfrak{q}, \mathfrak{p}]$, $\mathfrak{e} + \mathfrak{p}$, $\mathfrak{e} - \frac{1}{c}[\mathfrak{q}\,\mathfrak{m}]$, $\mathfrak{h} + \mathfrak{m}$ als Feldvektoren einzuführen und so die Gleichungen erheblich zu vereinfachen. Man erhält so Minkowskis Form der Feldgleichungen, welche mit der Form der Feldgleichungen für ruhende Körper bis auf das Glied $\mathfrak{q}\rho$ übereinstimmt. Wir thun dies aber nicht, weil die so einzuführenden Vektoren keine einfache physikalische Bedeutung besitzen. Was man bei ihrer Einführung an Einfachheit der Grundgleichungen gewinnt, das verliert man wieder ~~bei Aufst~~ dadurch, dass die Gleichungen, welche den Zusammenhang der Zustände der Körper mit den Feldvektoren angeben, komplizierter werden. Berücksichtigt man, dass die auf eine bewegte Elektrizitätseinheit wirkende Kraft gemäss §2 gleich $\mathfrak{e} + \frac{1}{c}[\mathfrak{q}, \mathfrak{h}]$ („elektromotorische Kraft") ist, und dass die auf die bewegte magnetische Ladungs-Einheit wirkende Kraft* gleich $\mathfrak{h} - \frac{1}{c}[\mathfrak{q}\,\mathfrak{e}]$ ist („magnetomotorische Kraft"), so erhält man ~~man~~ analog zu den Gleichungen (8')

$$\left.\begin{aligned} \mathfrak{p} &= (\varepsilon - 1)\left(\mathfrak{e} + \tfrac{1}{c}[\mathfrak{q}, \mathfrak{h}]\right) \\ \mathfrak{m} &= (\mu - 1)\left(\mathfrak{h} - \tfrac{1}{c}[\mathfrak{q}, \mathfrak{e}]\right) \\ \mathfrak{i} &= \lambda\left(\mathfrak{e} + \tfrac{1}{c}[\mathfrak{q}, \mathfrak{h}]\right) \end{aligned}\right\} (12)$$

---

* Man kann dies ~~entweder~~ leicht durch Erweiterung der in §2 gegebenen Betrachtungen beweisen, indem man auch eine magnetische Dichte und einen magnetischen Strom einführt.

Considered from the standpoint of the relativity theory, which will be developed later, these equations can hold only to a first degree of approximation, i.e., only if one neglects those terms in the equations that are multiplied by $\frac{|\mathfrak{q}|}{c}$ to the second or even higher power.[1] Since such terms possess practical meaning only in a very few cases, because the velocities occurring in practice are almost always very small in comparison with c, one can base one's calculations on these Lorentz equations in almost all of the cases that correspond to observable phenomena. From them and equations (12), for example, one can easily provide the theory for the previously mentioned experiments of Roentgen and Eichenwald, as well as for Wilson's experiment. We will not do so because we brought up these experiments only in order to support the basic assumptions of Lorentz's theory. But we must recall here an experiment of fundamental importance that provides a splendid confirmation of Lorentz's theory and thus constitutes one of the main pillars of the relativity theory, namely Fizeau's experiment.

Fizeau's experiment.[29] Using an interference method, Fizeau answered the following question experimentally: How does the velocity $\mathfrak{q}$ of a moving transparent body influence the propagation velocity (the phase velocity) of a light ray propagating through it? He found

$$V = V_0 + \mathfrak{q}_l \left(1 - \frac{1}{n^2}\right).$$

Here V_0 denotes the propagation velocity of the light in the body whose refraction index is n when it is at rest, and V the propagation velocity of the light in the body in the case where the latter moves in the direction of the propagation of the light with the velocity $\mathfrak{q}_l$. This can be derived from equations (Ib) and (12) in the following way.

For the X-axis of our coordinate system we choose the direction of the propagation of the light (and the velocity of the matter). The indicated equations can then be satisfied by the following assumptions

$$\begin{aligned} \mathfrak{e}_x &= 0 & \mathfrak{h}_x &= 0 \\ \mathfrak{e}_y &= f(x - Vt) & \mathfrak{h}_y &= 0 \\ \mathfrak{e}_z &= 0 & \mathfrak{h}_z &= \alpha f(x - Vt). \end{aligned}$$

In light of the first two of equations (12), it follows from this assumption that the polarization components $\mathfrak{p}_y$ and $\mathfrak{m}_z$ are different from zero, namely

$$\mathfrak{p}_y = (\varepsilon - 1)\left(1 - \alpha \frac{\mathfrak{q}_x}{c}\right) f$$

$$\mathfrak{m}_z = (\mu - 1)\left(\alpha - \frac{\mathfrak{q}_x}{c}\right) f$$

If one substitutes these expressions for the field strengths and polarizations (Ib), then the second and fourth of these equations as well as the first and third of equations (Ib) are satisfied if (while neglecting the terms of the order $\left(\frac{q_x}{c}\right)^2$) one adds the two conditions

$$c\alpha = V\varepsilon - V(\varepsilon - 1)\,\alpha\,\frac{\mathfrak{q}_x}{c} + \mathfrak{q}_x(\varepsilon - 1)$$

$$c = V\mu\alpha\,(\mu - 1)\,\frac{\mathfrak{q}_x}{c} + \mathfrak{q}_x(\mu - 1)\alpha$$

These two equations determine V and α. Eliminating α and keeping in mind the meaning of V_0 and n, one finds, after simple calculation, the equation that is to be proved

$$V = V_0 + \mathfrak{q}_x \left(1 - \frac{1}{n^2}\right).$$

[1] This has to do with the fact that, in the derivation of the equations, times, lengths, current strengths, polarizations have been treated as quantities defined independently of the state of motion of the reference system, which is inadmissible from the standpoint of the theory of relativity.

Vom Standpunkte der nachher zu entwickelnden Relativitätstheorie aus betrachtet können diese Gleichungen nur in erster Annäherung, d. h. nur dann gelten, wenn man die Glieder in den Gleichungen unberücksichtigt lässt, welche mit $\frac{|\mathfrak{q}|}{c}$ in der zweiten oder noch höheren Potenz multipliziert sind.[1] Da solchen Gliedern nur in sehr wenigen Fällen eine praktische Bedeutung zukommt, weil die praktisch vorkommenden Fälle der Geschwindigkeiten gegen c fast immer sehr klein sind, können diese Lorentz'schen Gleichungen in fast allen Fällen, denen beobachtbare Phänomene entsprechen, der Rechnung zu Grunde gelegt werden. Man kann z. B. aus ihnen und den Gleichungen (12) bequem die Theorie zu den schon erwähnten Experimenten von Röntgen und Eichenwald sowie zu dem von Wilson geben. Wir unterlassen dies, weil wir jene Experimente herangezogen, um die Grundannahmen der Lorentz'schen Theorie zu stützen. Eines fundamental wichtigen Experimentes aber müssen wir hier gedenken, das eine glänzende Bestätigung für die Lorentz'sche Theorie und damit eine der Hauptstützen der Relativitätstheorie bildet, nämlich des Experimentes von Fizeau.

Versuch von Fizeau. Fizeau beantwortete experimentell durch eine Interferenzmethode die folgende Frage. Wie beeinflusst die Geschwindigkeit $\mathfrak{q}$ eines bewegten durchsichtigen Körpers die Fortpflanzungs-Geschwindigkeit (Phasengeschwindigkeit) eines durch ihn sich fortpflanzenden Lichtstrahles? Er fand

$$V = V_0 + \mathfrak{q}_x\left(1 - \frac{1}{n^2}\right).$$

Dabei bedeutet V_0 die Fortpflanzungsgeschwindigkeit des Lichtes in dem Körper (vom Brechungsindex n), falls dieser ruht, V die Fortpflanzungsgeschwindigkeit des Lichtes im Körper in dem Falle, dass dieser in Richtung der Lichtfortpflanzung mit der Geschwindigkeit $\mathfrak{q}_x$ bewegt ist. Dies folgt man aus den Gleichungen (I b) und (12) in folgender Weise.

Wir wählen als X-Axe unseres Koordinatensystems die Richtung der Licht-Fortpflanzung (und Geschwindigkeit der Materie). Es lassen sich dann die genannten Gleichungen durch folgenden Ansatz befriedigen

$$\mathfrak{n}_x = 0 \qquad \mathfrak{d}_x = 0$$
$$\mathfrak{n}_y = f(x - Vt) \qquad \mathfrak{d}_y = 0$$
$$\mathfrak{n}_z = 0 \qquad \mathfrak{d}_z = \alpha f(x - Vt).$$

Aus ihm folgt vermöge der der beiden ersten der Gleichungen (12), dass nur die Polarisationskomponenten $\mathfrak{p}_y$ und $\mathfrak{m}_z$ von null verschieden sind, und zwar

$$\mathfrak{p}_y = (\varepsilon - 1)\left(1 - \alpha\frac{\mathfrak{q}_x}{c}\right) f$$
$$\mathfrak{m}_z = (\mu - 1)\left(\alpha - \frac{\mathfrak{q}_x}{c}\right) f$$

Setzt man diese Ausdrücke für die Feldstärken und Polarisationen (I b) ein, so sind die zweite und vierte dieser Gleichungen erfüllt, ebenso die erste und dritte der der Gleichungen (I b), wenn man (unter Vernachlässigung von Gliedern von der Ordnung $\left(\frac{\mathfrak{q}_x}{c}\right)^2$) die beiden Bedingungen hinzufügt

$$c\alpha = V\varepsilon - V(\varepsilon - 1)\alpha\frac{\mathfrak{q}_x}{c} + \mathfrak{q}_x(\varepsilon - 1)$$
$$c = V\mu\alpha - V(\mu - 1)\frac{\mathfrak{q}_x}{c} + \mathfrak{q}_x(\mu - 1)\alpha$$

Diese beiden Gleichungen bestimmen V und α. Durch Eliminieren von α findet man nach einfacher Rechnung unter Berücksichtigung der Bedeutung von V_0 und n die zu beweisende Gleichung

$$V = V_0 + \mathfrak{q}_x\left(1 - \frac{1}{n^2}\right).$$

---

1) Es hängt dies damit zusammen, dass bei der Ableitung der Gleichungen Zeiten, Längen, Stromstärke, Polarisationen als unabhängig vom Bewegungszustande des Bezugssystems definierte Grössen behandelt werden, was vom Standpunkte der Relativitätstheorie unzulässig ist.

SECTION 2

ELEMENTARY EXPOSITION OF THE FOUNDATIONS AND MOST IMPORTANT CONSEQUENCES OF THE RELATIVITY THEORY

§5. Principle of the Constancy of the Velocity of Light

As we have seen, Lorentz's theory, which was explained in Section 1 and which is the only viable refinement of Maxwell's theory so far, is based first of all on the assumption of the general validity of equations (I). But according to these equations, the propagation of light in empty space always occurs in accordance with equations of the form

$$\Delta\varphi - \frac{1}{c^2}\frac{\partial^2\varphi}{\partial t^2} = 0. \qquad \ldots (13)$$

For if one differentiates the first of equations (I) with respect to time and then eliminates $\dot{\mathfrak{h}}$ with the help of the third of equations (I), one obtains for the case when ρ vanishes, according to the mathematical rule ( ),[30]

$$\frac{1}{c}\ddot{\mathfrak{e}} = \text{curl } \dot{\mathfrak{h}} = c \text{ curl } (\text{curl } \mathfrak{e}) = -c\{-\Delta\mathfrak{e} + \text{grad } (\text{div } \mathfrak{e})\} = c\Delta\mathfrak{e}.$$

Thus, (13) holds for $\mathfrak{e}$ and its components and also, as can easily be shown, for $\mathfrak{h}$ and its components. But, as we know, equation (13) is solved by $f(x \pm ct)$, which form allows one to see that it represents an excitation that propagates in the form of a plane wave with velocity c. In virtue of equations (I), only electromagnetic waves propagating in vacuum with velocity c are actually possible in the case where $\rho = 0$.

Hence, in accordance with Lorentz's theory we can proclaim the following principle, which we call "the principle of the constancy of the velocity of light":

"There exists a coordinate system with respect to which every light ray propagates in vacuum with the velocity c."

This principle contains a far-reaching assertion. It asserts that the propagation velocity of light depends neither on the state of motion of the light source nor on the states of motion of the bodies surrounding the propagation space. The question as to what extent this principle can be considered certain is of fundamental significance for the theory of relativity. For the time being we will content ourselves with the realization that this principle is demanded by Lorentz's theory.

§6. The Principle of Relativity

The principle of relativity in classical mechanics. If the coordinate system K is chosen such that the equations of motion of material points are as simple as possible, then, if the customary notation is used, the equation of motion of the νth point is

$$m_\nu \frac{d^2\mathfrak{r}_\nu}{dt^2} = \mathfrak{K}_\nu, \qquad \ldots (14)$$

or in orthogonal components

$$m_\nu \frac{d^2x_\nu}{dt^2} = \mathfrak{K}_{\nu x} \text{ etc.} \qquad \ldots (14')$$

2. Abschnitt.

Elementare Darlegung der Grundlagen und wichtigsten Folgerungen der Relativitätstheorie.

§5 Prinzip von der Konstanz der ~~Vakuum~~ Lichtgeschwindigkeit. Die im 1. Abschnitt dargelegte Lorentz'sche Theorie, welche ~~sich stets~~ die bisher einzige ~~fast~~ lebensfähige Fortbildung von Maxwells Theorie ~~ist~~, beruht, wie wir gesehen haben, in erster Linie auf der Voraussetzung der allgemeinen Giltigkeit der Gleichungen (I). Nach diesen aber geschieht die Lichtfortpflanzung im leeren Raume stets nach den Gleichungen von der Form

$$\Delta\varphi - \frac{1}{c^2}\frac{\partial^2\varphi}{\partial t^2} = 0. \quad \ldots\ldots (13)$$

Differenziert man nämlich die erste der Gleichungen (I) nach der Zeit, und ersetzt man dann $\dot{\mathfrak{h}}$ mittelst der dritten der Gleichungen (I), so erhält man nach Rechnungsregel ( ) für den Fall dass ϱ verschwindet,

$$\frac{1}{c}\ddot{\mathfrak{e}} = \operatorname{curl}\dot{\mathfrak{h}} = -c\operatorname{curl}(\operatorname{curl}.\ \mathfrak{e}) = -c\left\{-\Delta\mathfrak{e} + \operatorname{grad}(\operatorname{div}\mathfrak{e})\right\} = -c\Delta\mathfrak{e}$$

Es gilt also (13) für $\mathfrak{e}$ und dessen Komponenten und, wie leicht zu zeigen, auch für $\mathfrak{h}$ und dessen Komponenten. Gleichung (13) wird aber bekanntlich durch $f(x \pm ct)$ gelöst, ~~Man ersieht~~ welche Form erkennen lässt, dass sie eine mit ~~Lichtg~~ der Geschwindigkeit c in Form einer ebenen Welle sich ausbreitende Erregung darstellt. Vermöge der Gleichungen (I) sind im Falle $\varrho = 0$ überhaupt nur solche elektromagnetische Wellen möglich, die sich im Vakuum mit der Geschwindigkeit c fortpflanzen.

Wir kommen daher im Einklang mit der H. A. Lorentz'schen Theorie den folgenden Grundsatz aufstellen, den wir „Prinzip von der Konstanz der Lichtgeschwindigkeit" nennen:

„Es gibt ein Koordinatensystem, inbezug auf welches sich jeder Vakuum-Lichtstrahl ~~im Vakuum~~ mit der Geschwindigkeit c fortpflanzt."

Dieser Satz ~~ist~~ enthält eine weitgehende Behauptung. Es wird durch ihn ~~ausgeschlossen~~ behauptet, dass die Fortpflanzungsgeschwindigkeit des Lichtes weder von dem Bewegungszustande der Lichtquelle noch vom Bewegungszustande der den Fortpflanzungsraum umgebenden Körper abhänge. Die Frage, inwieweit dieser Satz als gesichert gelten kann, ist ~~daher~~ von fundamentaler Bedeutung für die Relativitätstheorie. Wir wollen uns hier einstweilen mit der Einsicht begnügen, dass er durch die Lorentz'sche Theorie gefordert wird.

§6

Das Relativitätsprinzip.

Das Relativitätsprinzip in der klassischen Mechanik. Wählt man das Koordinatensystem K so, dass die Bewegungsgleichungen der materiellen Punkte möglichst einfach werden, so lautet bei üblicher Bezeichnungsweise die Bewegungsgleichung des ν, ten Punktes

$$m_\nu \frac{d^2\mathfrak{r}_\nu}{dt^2} = \mathfrak{K}_\nu, \quad \ldots\ldots (14)$$

oder in ~~Koo~~ rechtwinkligen Komponenten

$$m_\nu \frac{d^2 x_\nu}{dt^2} = \mathfrak{K}_{\nu x} \text{ etc.} \quad \ldots (14')$$

If the system is a complete one, i.e., if only masses that belong to the system act upon its material points, then the force components $\mathfrak{K}_{\nu x}$ etc. can always be represented according to the customary assumptions as functions of the differences of the same kind of coordinates of the points, i.e., as functions of the differences $x_\nu - x_1, x_\nu - x_2, \ldots z_\nu - z_l$.

If one now introduces a new coordinate system K' whose axes are continually parallel to those of K but whose origin moves with constant speed along a straight line, *then one obtains equations of motion that read exactly the same as the original ones.* We can assume without loss of generality that the straight line along which the coordinate origin glides is the X-axis of K. In that case, according to the customary kinematics, we have the transformation equations

$$\left.\begin{aligned} x' &= x - vt \\ y' &= y \\ z' &= z \\ t' &= t \end{aligned}\right\} \qquad \text{(II)}$$

where x', y', z' are the coordinates in the new system. The equation $t' = t$ has been appended so as to permit the designation of all spatial and temporal determinations by primed letters in the new system. Since it follows from the first of equations (II) that

$$x_\mu' - x_\nu' = x_\mu - x_\nu \,,$$

and from the third and fourth of equations II that

$$\frac{dx'}{dt'} = \frac{dx}{dt} - v$$

$$\text{and} \quad \frac{d^2x'^2}{dt'} = \frac{d^2x}{dt^2} \,,$$

then the assertion has been proved, if one considers that the y- and z-coordinates have not been changed at all. If, along with Laue, we call a coordinate system for which the equations (14) hold, with the mentioned restrictions, a "justified" system,[31] we can assert the following propositions in the sense of classical mechanics

> "Every coordinate system that is in uniform translational motion relative to a justified system is again a justified system. The equations of motion of any (mechanical)* system are the same with respect to all such justified systems."

We will call this proposition the "relativity principle."

Relativity principle and experience. Astronomy has led us to conceive of a coordinate system whose origin is at the center of gravity of the solar system and whose axes are directed toward specific points of the heaven of the fixed stars as a "justified" system. Relative to this system, Earth has a rotational motion and a translational motion. The latter corresponds to its orbital motion around the sun; the velocity of

* This restriction to the *mechanical* laws shall be dropped hereafter.

Ist das System ein vollständiges, d. h. wirken auf die materiellen Punkte nur Massen ein, welche zum System gehören, dann lassen sich die ~~Kräfte R_ν~~ Kraftkomponenten $R_{\nu x}$ etc. nach den üblichen Voraussetzungen stets als Funktionen der Differenzen gleichartiger Koordinaten der Punkte, also als Funktionen der Differenzen $x_\nu - x_1, x_\nu - x_2; \dots z_\nu - z_l$ darstellen.

Führt man nun ein neues Koordinatensystem K' ein, dessen ~~Anfangspunkt~~ Achsen dauernd mit denjenigen von K parallel sind, dessen Anfangspunkt sich aber auf einer Geraden mit konstanter Geschwindigkeit bewegt, so erhält man Bewegungsgleichungen die genau gleich lauten wie die ursprünglichen. Wir können ohne Beschränkung der Allgemeinheit annehmen, dass die Gerade, auf welcher der ~~der~~ Koordinatenursprung gleitet, die X-Achse von K sei. Dann gelten nach der uns geläufigen Kinematik die Transformationsgleichungen

$$\left.\begin{aligned} x' &= x - vt \\ y' &= y \\ z' &= z \\ t' &= t \end{aligned}\right\} \quad (II)$$

$x' y' z'$ sind hiebei die Koordinaten im neuen System. Die Gleichung $t' = t$ ist beigefügt, um im neuen System alle räumlichen und zeitlichen Angaben mit gestrichelten Buchstaben bezeichnen zu können. Da aus der ersten der Gleichungen (II) hervorgeht, dass

$$x'_\mu - x'_\nu = x_\mu - x_\nu,$$

und da aus der ersten und vierten der Gleichungen II folgt

$$\frac{dx'}{dt'} = \frac{dx}{dt} - v$$

und

$$\frac{d^2x'}{dt'^2} = \frac{d^2x}{dt^2},$$

so ist die Behauptung erwiesen, wenn man bedenkt, dass an den y- und z-Koordinaten überhaupt nichts geändert wird. Nennen wir mit Laue ein Koordinatensystem, für welches die Gleichungen (14) mit den angegebenen Einschränkungen gelten, ein „berechtigtes" System, so können wir im Sinne der klassischen Mechanik die Sätze aussprechen:

„Jedes relativ zu einem berechtigten System in gleichförmiger Translationsbewegung befindliches Koordinatensystem ist wieder ein berechtigtes System. Inbezug auf alle derartigen berechtigten Systeme sind die Bewegungsgesetze irgend eines (mechanischen)* Systems die nämlichen."

Diesen Satz wollen wir kurz als das „Relativitätsprinzip" bezeichnen.

Relativitätsprinzip und Erfahrung. Die Astronomie hat dazu geführt, ~~[illegible]~~ ein Koordinatensystem, dessen Ursprung im Schwerpunkt des Sonnensystems, und dessen Achsen nach bestimmten Punkten des Fixsternhimmels gerichtet sind, als ein „berechtigtes" System aufzufassen. Die Erde hat relativ zu diesem System eine Drehbewegung und eine fortschreitende Bewegung. Die letztere entspricht ihrer Bahnbewegung um die Sonne; die Geschwindigkeit dieser Bewegung ist ungefähr 30 km pro Sekunde. Sehen wir

* Diese Beschränkung auf die mechanischen Gesetze wird gleich nachher fallen gelassen.

this motion is approximately 30 km per second. Thus, if we disregard the rotational motion of the Earth and the acceleration that the Earth experiences in its orbital motion, then a coordinate system at rest relative to the Earth is to be conceived as a system possessing a uniform translational velocity of about 30 km/sec with respect to a "justified" system. If the relativity principle were not valid, then it would have to be possible to produce a proof of that motion by means of mechanical experiments carried out on the Earth. In particular, it would have to be expected that the direction of that motion would be somehow privileged with respect to directions perpendicular to it. Obviously, such a physical

One can go even further. If the mentioned coordinate system at rest relative to the solar system (or some other coordinate system in a specific state of motion) would be privileged with respects to systems in uniform translational motion relative to it as regards *mechanical laws*, in that these laws would be especially simple with respect to that system, then an analogous behavior would also be expected as regards the rest of the physical laws. The question thus arises as to whether it is possible to demonstrate an influence of the Earth's motion, or of its changing direction with the time of year, on some physical states or processes. It is well known that all such efforts to prove such an influence of Earth's translational motion by physical methods have failed.

anisotropy of our terrestrial observational spaces resulting from the Earth's motion would not be restricted to purely mechanical phenomena. In general, the question arises whether the physical equilibria and processes that we observe in our laboratories depend on the orientation of the whole system relative to a (seasonally dependent) direction. Many such experiments have been undertaken, among them the fundamentally important one by Michelson and Morley, <which will be dealt with soon.>[32] familiarity with which can surely be taken for granted here. An influence of the Earth's motion could not be proved anywhere. This makes the just propounded relativity principle an almost indisputable fact, which seems to be valid not only in mechanics but in all branches of physics. Still, one has to add one restriction: the relativity principle could also be violated if no spatial anisotropy occurs, but only the *magnitude* of the Earth's velocity were to play a role. Thus, for example, the ratio of the length of a specific solid body to the wavelength of a particular spectral line, investigated on the Earth, might depend on the magnitude of the velocity of the Earth's motion relative to some fundamental system. After what has been learned regarding the influence of the *direction* of the Earth's motion on terrestrial experiments, everybody will probably have the courage to predict a negative result of such efforts. I believe therefore that the validity of the relativity principle can hardly be doubted any longer.

also ab von der Drehbewegung der Erde und von der Beschleunigung, welche die Erde bei ihrer Bahnbewegung erfährt, so ist ein relativ zur Erde ruhendes Koordinatensystem als ein solches aufzufassen, welches inbezug auf ein „berechtigtes" Bezugssystem eine gleichmässige Translations-Geschwindigkeit von etwa 30 km/sek. besitzt. Wenn das Relativitätsprinzip nicht gültig wäre, so sollte es möglich sein durch auf der Erde ausgeführte mechanische ~~Versuche~~ Versuche den Nachweis jener Bewegung zu erbringen. Insbesondere sollte erwartet werden, dass die Richtung jener Bewegung gegenüber den dazu senkrechten Richtungen irgendwie ausgezeichnet wäre. Eine derartige physikalische

~~Man kann noch weiter gehen. Wenn das erwähnte, relativ zum Sonnensystem ruhende Koordinatensystem (oder irgend ein anderes Koordinatensystem von bestimmtem Bewegungszustande) inbezug auf die mechanischen Gesetze ~~Gesetze~~ gegenüber den relativ zu ihm in gleichförmiger Translationsbewegung befindlichen Systemen dadurch ausgezeichnet wäre, dass jene Gesetze inbezug auf jenes System besonders einfache wären, so wäre ein analoges Verhalten auch bezüglich der übrigen physikalischen Gesetze zu erwarten. Es ergibt sich also die ~~Fragestellung~~, ob sich ein Einfluss der Erdbewegung bzw. der mit der Jahreszeit sich ändernden Richtung der Erdbewegung auf irgend welche ~~auf~~ physikalische Zustände oder Vorgänge konstatieren lässt. Es ist wohlbekannt, dass alle derartigen ~~Versuche, die~~ Bemühungen einen derartigen Einfluss der Translationsbewegung der Erde physikalisch nachzuweisen, gescheitert sind.~~

~~Von dieser Erwägung ausgehend und~~

Anisotropie unserer irdischen Beobachtungsräume infolge der Erdbewegung wäre offenbar nicht auf rein mechanische Phänomene beschränkt. Es entsteht ganz allgemein die Frage, ob physikalische Gleichgewichte und Vorgänge, die wir in unseren Laboratorien beobachten, von der Orientierung des ganzen Systems relativ zu einer (von der Jahreszeit abhängigen) Richtung ~~bezw~~ abhängen. Es wurden viele solche Versuche unternommen, darunter der fundamental wichtige von Michelson und Morley, den ich hier wohl als bekannt voraussetzen dürfen ~~von dem bald noch die Rede sein wird~~. Ein Einfluss der Erdbewegung konnte nirgends nachgewiesen werden. Dadurch wird das vorhin ausgesprochene Relativitätsprinzip ~~als~~ beinahe zu einer unbezweifelbaren Thatsache, und zwar scheint dasselbe nicht nur in der Mechanik, sondern in allen Zweigen der Physik ~~Gültigkeit zu besitzen~~ gültig zu sein. Eine Beschränkung muss immerhin hinzugefügt werden, das Relativitätsprinzip könnte auch in der Weise verletzt sein, dass keine ~~Betrag~~ Anisotropie des Raumes eintritt sondern nur der Betrag der Erd-Geschwindigkeit eine Rolle spielte. Es könnte also z. B. das Verhältnis der ~~die~~ Länge eines bestimmten festen Körpers, ~~gemessen durch die Angabe~~ zur Wellenlänge einer bestimmten Spektrallinie, auf der Erde untersucht, vom Betrage der Geschwindigkeit der Erdbewegung zu irgend einem Fundamentalsystem abhängen. Nach den bezüglich des Einflusses der Richtung der Erdbewegung auf die terrestrischen Experimente gemachten Erfahrungen wird wohl jeder den Mut haben, ein negatives Resultat solcher Bemühungen zu prophezeihen. Ich glaube deshalb, dass an der Gültigkeit des Relativitätsprinzipes kaum mehr zu zweifeln ist.

§7. Apparent Incompatibility of the Principle of the Constancy of the Speed of Light with the Relativity Principle

Not nearly so many experiments would have been undertaken to prove the influence of Earth's motion on terrestrial experiments had Lorentz's theory, which I described earlier, not seemed incompatible with the relativity principle. For we have seen in §5 that, according to Lorentz's theory, there exists a reference system K in which every light ray propagates in vacuum with the velocity c. It would appear that a system in uniform translational motion with respect to K could not have this property. For if again we choose K' such that its origin glides along the X-axis of K with velocity v, and if we choose a light ray that propagates along the X-axis of K with the velocity c with respect to K, then, to all appearances (according to the law of the parallelogram of velocities), as well as according to the transformation equations (II), this same light ray propagates along the X'-axis with velocity $c - v$ with respect to K'. Because owing to equations (II), the equation $x = ct$ is equivalent to the equation $x' + vt' = ct'$. From this it would follow that a law of nature, namely, the principle of the constancy of the velocity of light, is indeed valid with respect to K but not with respect to K'. Thus the principle of the constancy of the velocity of light (in vacuum) seems to be incompatible with the relativity principle.

This state of affairs compels us to question again the correctness of the principle of the constancy of the velocity of light and, therewith, the fundamental equations (I) of Lorentz's theory. We must entertain the possibility that the principle of the constancy of the velocity of light perhaps cannot possess strict validity despite the successes of Lorentz's theory; but we would have to keep the relativity principle all the same, and also the kinematic equations II, i.e., the law of the parallelogram of velocities.

If we take this standpoint, then Fizeau's experiment forces us to conclude that in a transparent medium (neglecting dispersion) various velocities of light must be possible. For if the velocity of light in the medium as assessed by an observer moving with the medium were always equal to V_0, then, according to the parallelogram law of velocities, the velocity of light as assessed by an observer not moving with the medium would be $V_0 + \mathfrak{q}_l$, which contradicts experimental findings. We are thus forced to assume that, in Fizeau's experiment, the velocity of light relative to the moving medium is, in this medium, different from what it would be in the same medium if the latter were at rest. Since this must also hold if the medium is optically inactive,[33] i.e., if $n = 1$, we can

§7.

Scheinbare Unvereinbarkeit des Prinzipes der Konstanz der Lichtgeschwindigkeit
mit dem Relativitätsprinzip.

Es wären wohl kaum so zahlreiche Versuche unternommen worden
einen Einfluss der Erdbewegung auf terrestrische Experimente nachzuweisen, wenn nicht die geschilderte Lorentz'sche Theorie mit dem
Relativitätsprinzip unvereinbar zu sein schiene. Wir haben nämlich
in §5 gesehen, dass es nach der Lorentz'schen Theorie ein Bezugssystem K gibt,
in welchem sich jeder Vakuum-Lichtstrahl mit der Geschwindigkeit c fortpflanzt. Es hat nun den Anschein, wie wenn ein zu K in gleichförmiger
Translationsbewegung befindliches System diese Eigenschaft nicht haben
könnte. Wählen wir nämlich K' wieder ~~so bewegte~~ derart, dass sein Ursprung
auf der X Achse von K mit der Geschwindigkeit v gleitet, und wählen
einen Lichtstrahl, der inbezug auf K längs der X-Axe von K mit der Geschwindigkeit c sich ausbreitet, so breitet sich inbezug auf K'
derselbe Lichtstrahl längs der X' Achse ~~mit~~ dem Anscheine nach
(nach dem Satz vom Parallelogramm der Geschwindigkeiten) sowie
gemäss den Verwandlungsgleichungen $(\underline{II})$ mit der Geschwindigkeit
$c-v$ aus. Denn die Gleichung $x=ct$ ist vermöge der Gleichungen $(\underline{II})$ mit
der Gleichung $x'+vt'=ct'$ gleichbedeutend. Es würde daraus folgen,
dass ein Naturgesetz, nämlich das Prinzip von der Konstanz der
Lichtgeschwindigkeit, wohl inbezug auf K, nicht aber inbezug
auf K' gültig wäre. Das Prinzip von der Konstanz der (Vakuum-) Lichtgeschwindigkeit scheint also mit dem Relativitätsprinzip unvereinbar zu sein.

Diese Sachlage drängt uns dazu, an der Richtigkeit des Prinzips
von der Konstanz der Lichtgeschwindigkeit, und damit an den Grundgleichungen $(\underline{I})$ der Lorentz'schen Theorie abermals zu zweifeln. Wir müssen
die Möglichkeit in Erwägung ziehen, dass trotz der Erfolge der Lorentz'schen
Theorie das Prinzip von der Konstanz der Lichtgeschwindigkeit ~~trotz dem~~
vielleicht keine strenge Gültigkeit besitzen könnte; dabei hätten wir
jedoch am Relativitätsprinzip festzuhalten und ebenso an den kinema-
tischen Gleichungen $\underline{II}$, d. h. an dem Gesetz vom Parallelogramm der
Geschwindigkeiten.

Wenn wir uns auf diesen Standpunkt stellen, drängt uns das
Fizeau'sche Experiment dazu, dass (abgesehen von der Dispersion) in
einem durchsichtigen Medium verschiedene Lichtgeschwindigkeiten
möglich sein müssten. Wäre nämlich die Lichtgeschwindigkeit im
dem Medium, von einem mit dem Medium bewegten Beobachter aus
beurteilt, ~~glei~~ stets gleich V_0, so wäre nach dem Parallelogramm-Gesetz
der Geschwindigkeiten die Lichtgeschwindigkeit, von dem nicht mit dem
Medium bewegten Beobachter aus beurteilt, gleich $V_0 + v$, was mit
dem experimentellen Befund im Widerspruch ist. Wir werden also
~~immer~~ zu der Annahme gedrängt, dass relativ zum bewegten Medium beim Fizeau'schen
Versuch die Lichtgeschwindigkeit in diesem eine andere sei als in demselben Medium, falls dieses ruht. Da dies auch dann gelten muss,
wenn das Medium optisch unwirksam ist, d. h. wenn $n=1$ ist, so kommen

Strictly speaking, however, we have only learned that the following three things are incompatible with one another:

(a) the relativity principle

(b) the principle of the constancy of the velocity of light (Lorentz's theory)

(c) the transformation equations (II), or the law of the parallelogram of velocities

One arrives at the theory that is now called "the theory of relativity" by keeping (a) and (b) but rejecting (c).[34] In what follows, it will become evident that it is *possible* to proceed in this way. But whoever has not yet studied the electrodynamics of moving bodies thoroughly will surely be inclined to keep (a) and (c) but give up (b). Therefore we will first draw several consequences starting out from the latter standpoint in order to justify in that way the standpoint of the theory of relativity.—

If we consider the arrangement in Fizeau's experiment in principle, then we have to distinguish the following objects:

(1) the light source (*L*)

(2) the medium traversed by the light (*M*)

(3) the Earth with all the remaining objects (*E*).[35]

We can characterize the state of affairs in Fizeau's experiment adequately in the following way: *L* and *E* are at rest with respect to the coordinate system *K*. *M* is traversed by the light (velocity of light *V*) in the direction of the x-axis and is moving with the velocity $\mathfrak{q}_l$. According to (a), this state of affairs is equivalent to the following one: *M* is at rest; *L* and *E* possess the velocity $-\mathfrak{q}_l$ along the *X*-axis. According to Fizeau's result and c), the velocity of light V' that obtains with respect to *M* is in this case equal to

$$V' = \left\{ V_0 + \mathfrak{q}_l\left(1 - \frac{1}{n^2}\right)\right\} - \mathfrak{q}_l = V_0 - \frac{\mathfrak{q}_l}{n^2},$$

where V_0 is the velocity of light in *M* if *L M* and *E* are at rest relative to one another. Thus, according to this conception, the circumstance that *L* and *E* are moving with the velocity $-\mathfrak{q}_l$ with respect to *M* changes the velocity of light in *M* by $\left(-\frac{\mathfrak{q}_l}{n^2}\right)$. One can hold either the motion of *E* or the motion of *L* responsible for this change $V' - V_0$. (We call these two possibilities "case I" and "case II").

Case I leads to great complications. For the influence of the motion of *E* with respect to *M* on the velocity of light in *M* must diminish as the mass of *E* is made smaller, and that of *M* larger.* In the case of a real experiment, the situation would have to be conceived in such a way that the effective mass of *E* outweighs that of *M* to such an extent that it alone has to be considered. But since there are no clues whatsoever regarding the law according to which moving masses could modify the velocity of light in surrounding spaces,[36] this conception gives free rein to arbitrary choice. In any case, the mass-effect would have to diminish drastically with the distance because otherwise one would run into conflict with the result of Michelson's experiment.[46]

* Fizeau's experiment—carried out far from the Earth in free space—would have to yield the result $V = V_0 + \mathfrak{q}_l$.— It should also be noted that the explanation of the aberration from this standpoint meets with great difficulties.

Genau genommen haben wir aber nur eingesehen, dass folgende drei Dinge miteinander nicht vereinbar sind

a) das Relativitätsprinzip

b) das Prinzip der Konstanz der Lichtgeschwindigkeit (Lorentz'sche Theorie)

c) die Transformationsgleichungen (II), bezw. das Gesetz vom Parallelogramm der Geschwindigkeiten

Zu der heute als „Relativitätstheorie" bezeichneten Theorie gelangt man, indem man a) und b) beibehält, c) aber verwirft. Dass diese Art des Vorgehens möglich ist, wird sich im Folgenden zeigen. Jeder, der sich noch nicht eingehend mit der Elektrodynamik bewegter Körper befasst hat, wird aber sicherlich geneigt sein a) und c) beizubehalten, b) aber aufzugeben. Deshalb wollen wir zunächst ~~uns~~ von dem letzteren Standpunkte ausgehend einige Konsequenzen ziehen, um den Standpunkt der Relativitätstheorie dadurch zu rechtfertigen. –

Fassen wir die beim Fizeau'schen Versuch vorhandene Anordnung prinzipiell auf, so haben wir folgende Gegenstände zu unterscheiden

1) die Lichtquelle (L)

2) das vom Lichte durchsetzte Medium (M)

3) die Erde mit allen übrigen Gegenständen (E).

Den Sachverhalt beim Fizeau'schen Versuch können wir hinreichend so charakterisieren: L und E ruhen, ~~M ist bezüglich~~ inbezug auf das Koordinatensystem K M ist in Richtung der x-Achse durchstrahlt ~~und~~ (Lichtgeschwindigkeit V) und mit der Geschwindigkeit v_e bewegt. Diesem Sachverhalt ist nach a) der folgende gleichwertig: M ruht; L und E haben längs der x-Achse die Geschwindigkeit $-v_e$. Die dabei bezüglich M auftretende Lichtgeschwindigkeit V' ist nach dem ~~Befunde~~ Resultat Fizeaus und nach c) gleich

$$V' = \left\{ V_0 + v_e\left(1 - \frac{1}{n^2}\right) \right\} - v_e = V_0 - \frac{v_e}{n^2},$$

wobei V_0 die Lichtgeschwindigkeit in M ist für den Fall, dass L M und E relativ zu einander ruhen. Der Umstand, dass L und E bezüglich M ~~kein~~ mit der Geschwindigkeit $-v_e$ bewegt sind, ändert also bei dieser Auffassung die Lichtgeschwindigkeit in M um $\left(-\frac{v_e}{n^2}\right)$. Man kann nun für diese Aenderung $V' - V_0$ entweder die Bewegung von E oder die Bewegung von L verantwortlich machen (Diese beiden Möglichkeiten bezeichnen wir als „Fall I" und „Fall II").

Fall I führt zu grossen Komplikationen. Denn es müsste der Einfluss der Bewegung von E gegen M ~~desto~~ auf die Lichtgeschwindigkeit in M desto kleiner werden, je kleiner die Masse (?) von E und je grösser diejenige von M gemacht wird.* Im Falle des wirklichen Experimentes wäre die Sachlage so aufzufassen, dass die wirksame Masse von E gegenüber derjenigen von M so überwiegt, dass sie allein in Betracht kommt. Da es aber gar keinen Anhaltspunkt dafür gibt, nach welchem Gesetze bewegte Massen die Lichtgeschwindigkeit im umgebenden Raume modifizieren könnten, ist bei dieser Auffassung der Willkür freier Spielraum gelassen. Jedenfalls müsste jene Massen-Wirkung mit der Entfernung kräftig abnehmen, da man sonst mit dem Ergebnis des Versuches von Michelson in Konflikt gerät.

* Es müsste der Fizeau'sche Versuch – fern von der Erde im freien Weltraum ausgeführt – das Resultat $V = V_0 + v_e$ ergeben. – Es sei auch erwähnt, dass die Erklärung der Aberration von diesem Standpunkte aus den grössten Schwierigkeiten begegnet.

In case I,[38] the velocity of light would depend on the state of motion of the light source.* Light rays of most diverse propagation velocities could be present at one and the same location at the same time. The physical properties of the light would not be determined, then, by the frequency alone, since the wavelength and the frequency would be independent of each other; the light originating from stars that are moving relative to us would have to be physically distinguishable from the light originating from light sources at rest. Experience has yielded nothing of the sort. The most convincing argument against this point of view, which had been advocated by Ritz, has been put forward by the Dutch astronomer Pexider.[39] The individual stars of stellar systems (double stars) must have sent us light of different velocities during different epochs of their orbits. Thus, the time of propagation of the light from double stars to us would be different for different epochs. The temporal sequence of the epochs as traced by us with the help of the Doppler principle would be different from that in reality; a simple calculation shows that, indeed, if the underlying hypothesis were borne out by the facts, the indicated influence would have to be so considerable that it would have been absolutely impossible for the astronomers to miss it. The untenability of this conception can surely be viewed as being definitively proved.

In case II,[40] the motion of the surrounding matter would have an influence on the propagation velocity of light in stationary media. The influence of the motion of the matter traversed by light on the result of Fizeau's experiment would be attributable to the fact that the *Earth* is moving relative to the medium traversed by the light. Then one would have to assume that Fizeau's experiment would turn out differently if it were conducted on a sufficiently small celestial body. Even though this conception cannot be viewed as definitively disproved, still its extraordinary complexity already makes its correctness improbable. It is hardly plausible that the velocity of light in a medium would depend on the state of motion of the surrounding bodies, but would do so to a noticeable degree only if these bodies possessed a very large mass relative to the body traversed by the light. Also, it has not proved possible thus far to base a serviceable theory on this assumption.

On the other hand, since the Maxwell-Lorentz theory requires the principle of the constancy of the (vacuum) velocity of light, there is every reason to stick with it. It turns out that the result of the Fizeau experiment can be derived quantitatively from (a) and (b) without introducing the more special assumptions of the Maxwell-Lorentz theory. Laue was the first to draw attention to this important circumstance, to which we will return in §11.[41]

---

* The above formula shows that in the case of propagation of light in the vacuum ($n = 1$), the motion velocity of the source of light would simply have to be added to the velocity of the propagation of light of a stationary light source.

Im Falle I würde die Lichtgeschwindigkeit ~~eines vo~~ vom Bewegungszustande
der Lichtquelle abhängen. Es ~~würden~~ könnten an einem und demselben Orte zu derselben Zeit
Lichtstrahlen ~~vorhanden sein~~ der verschiedensten Ausbreitungs-Geschwindigkeiten
vorhanden sein. ~~Um ferner dem Fizeau'schen Experiment, wie es thatsächlich ausgeführt wurde, bei dieser Auf-~~
~~fassung gerecht zu werden, müsste man noch dazu annehmen, dass~~
~~die Geschwindigkeit eines Lichtstrahles~~ Es wären dann die physikalischen
Eigenschaften des Lichtes nicht durch die Frequenz allein bestimmt, indem
Wellenlänge und Frequenz voneinander unabhängig wären; das Licht,
welches von relativ zu uns bewegten Sternen herrührt, würde physikalisch
unterscheidbar sein müssen von dem Lichte, das von ruhenden Lichtquellen
herstammt. ~~Die überzeugendste Erfahr~~ Nichts von dem hat die Erfahrung
ergeben. Den überzeugendsten Grund gegen diese, von Ritz vertretene
Auffassungsweise hat der holländische Astronom Desitter geltend
gemacht. Die ~~Komponenten~~ einzelnen Sterne von Sternsystemen (Doppel-
sternen) müssten in verschiedenen Epochen ihres Umlaufes Licht mit
verschiedenen Geschwindigkeiten zu uns senden. Es würde also die
Zeit der Lichtausbreitung vom Doppelstern bis zu uns für die verschiedenen
Epochen verschieden sein. Die zeitliche Aufeinanderfolge (der Epochen, wie sie) bei uns
mittelst des Doppler'schen Prinzips ~~verfolgten Epochen~~ wird, wäre
also eine andere ~~wie diejenige~~ als in Wirklichkeit, und zwar zeigt
eine einfache Rechnung, dass jener Einfluss ein ~~ganz~~ derart erheblicher
sein müsste, dass er den Astronomen unmöglich hätte entgehen
können, wenn die zugrunde gelegte Hypothese den Thatsachen entspräche.
~~Diese~~ Die Unhaltbarkeit dieser Auffassung kann ~~also~~ wohl als ~~unhaltbar~~ endgültig erwiesen betrachtet werden.

Im Falle II hätte die Bewegung der umliegenden Materie einen
Einfluss auf die Ausbreitungsgeschwindigkeit des Lichtes in ruhenden
Medien. ~~Der~~ Der Einfluss der Bewegung der vom Licht durchsetzten Materie auf das Ergebnis des Fizeau'schen Versuches wäre
darauf zurückzuführen, dass die Erde relativ zum durchstrahlten Medium
sich in Bewegung befindet. Man müsste dann annehmen, dass das Experiment Fizeaus anders ausfallen würde, wenn es auf einem hinreichend
kleinen Himmelskörper ausgeführt würde. Wenn auch diese Auffassung
nicht als endgültig widerlegt gelten kann, so ist ~~sie~~ ihre Richtigkeit doch schon
wegen ihrer ausserordentlichen Kompliziertheit unwahrscheinlich.
Es ist kaum glaublich, dass die Lichtgeschwindigkeit in einem Medium vom
Bewegungszustand der umgebenden ~~Mass~~ Körper abhänge, aber nur dann
in merklichem Masse, wenn jene Körper gegenüber dem durchstrahlten eine sehr grosse Masse besitzen.
Es ist auch bisher nicht gelungen, auf diese Annahme eine brauchbare
Theorie zu gründen.

Da andererseits das Prinzip von der Konstanz der (Vakuum-)
Lichtgeschwindigkeit von der Maxwell-Lorenz-schen Theorie gefordert
wird, haben wir ~~beim heutigen Stande der Erfahrung alle~~ alle Ursache, an
ihm festzuhalten. Es zeigt sich, dass das Ergebnis des Fizeau'schen Versuches sich
quantitativ (aus a) und b) herleiten lässt, ohne Einführung der speziellen ~~Folge~~ Voraussetzungen der Maxwell-
Lorentz'schen Theorie. Auf diesen wichtigen Umstand, auf den wir in § 11 zurückkommen
werden, hat zuerst Laue aufmerksam gemacht*.

---

*Die obige Formel zeigt, dass sich ~~t~~ im Falle der Lichtausbreitung im Vakuum ($n=1$)
die Geschwindigkeit der Bewegung der Lichtquelle einfach zur Vakuum-Licht-
Geschwindigkeit einer ruhenden Lichtquelle addieren müsste.

In case II the velocity of light in M depends on the velocity of motion of the light source with respect to M (Ritz <and Ehrenfest>).[42] This being so, light rays of all possible propagation velocities, arbitrarily small or arbitrarily large, could occur in M. Intensity, color, and polarization state would not suffice to define a plane light wave; one would have also to add the determinative element of *velocity*, which, however, should not make itself felt in any effects of the first order (which would be proportional to the first power of the velocity of the light source). For the light coming from stars that are in motion relative to the Earth has—as far as our experience extends—the same properties as the light from terrestrial sources of light. To do justice to that, one is forced to make the most peculiar assumptions if one pursues this point of view, as for example the following: if light of velocity $c + v$ strikes a mirror perpendicularly, then the reflected light has the velocity $c - v$. These complications make it seem understandable why it has not proved possible so far to set up differential equations and boundary conditions that would do justice to this conception.— [43]

If, in conjunction with such reflections, one considers in how simple a fashion the theory of relativity interprets Fizeau's result ( ), one can hardly avoid the impression that, given the present state of physical experience, this theory is to be viewed as the most natural one.[44]

§8. The Physical Meaning of Spatial and Temporal Determinations

Prompted by the difficulties described in the previous §, we now seek to solve the dilemma that consists in the incompatibility of postulates (a), (b), and (c) of the previous § by relinquishing (c) but keeping the empirically supported postulates (a) and (b). We realize that the postulate (c) can be given up, i.e., that it is not logically necessary, by investigating the meaning that the spatial and temporal determinations have in physics.

Physical meaning of spatial determinations. The propositions of Euclidean geometry acquire a physical content through our assumption that there exist objects that possess the properties of the basic structures of Euclidean geometry. We assume that appropriately constructed edges of solid bodies not subjected to external influences have the definitional properties that straight lines have (material straight line), and that the part of a material straight line between two marked material points has the properties of a segment. Then the propositions of geometry turn into propositions concerning the arrangements of material straight lines and segments that are possible[46] when these structures are at relative rest.* By virtue of its properties, every material straight line can be extended. By repeatedly laying a shorter segment on a longer one and counting these operations, one can measure the latter by means of the former; i.e., the length of the latter can be expressed by a number if the former (the measuring rod) is regarded as being given once and for all. The position of a material point on a material straight line on which a fundamental point is given once and for all can be defined by its distance from the latter, that is, by a number (coordinate).

* Actually, this concept is wooly (rotation).[45] But I would like to avoid dwelling on things whose detailed consideration is not necessary for the goal to be pursued here.

Bei Fall II hängt die Lichtgeschwindigkeit in M von der Bewegungsgeschwindigkeit der Lichtquelle gegen M ab (Ritz ~~und Ehrenfest~~). Es könnte dann in M Lichtstrahlen von allen möglichen, beliebig kleinen und beliebig grossen Ausbreitungsgeschwindigkeiten geben. Eine ebene Lichtwelle wäre durch Intensität, Farbe und Polarisationszustand noch nicht definiert; man müsste noch das Bestimmungsstück der Geschwindigkeit hinzufügen, das aber bei allen Effekten erster Ordnung (die der ersten Potenz der Geschwindigkeit der Lichtquelle proportional wären) sich nicht geltend machen dürfte. Denn das Licht, welches von relativ zur Erde bewegten Sternen kommt, hat – soweit unsere Erfahrungen reichen – dieselben Eigenschaften wie das Licht irdischer Lichtquellen. Um dem gerecht zu werden wird man bei Verfolgung dieses Standpunktes zu den sonderbarsten Annahmen gezwungen, z. B. zu der folgenden: Fällt Licht von der Geschwindigkeit $c+v$ senkrecht auf einen Spiegel, so hat das reflektierte Licht die Geschwindigkeit $c-v$. Diese Komplikationen lassen es begreiflich erscheinen, dass es bisher nicht gelungen ist, Differentialgleichungen und Grenzbedingungen aufzustellen, welche dieser Auffassung gerecht werden.–

Erwägt man im Zusammenhange mit solchen Überlegungen, auf wie einfache Weise die Relativitätstheorie Fizeau's Ergebnis deutet (   ), so wird man sich des Eindruckes kaum erwehren können, dass diese Theorie beim heutigen Stande der physikalischen Erfahrung als die natürlichste anzusehen ist.

§8.

Physikalische Bedeutung räumlicher und zeitlicher Angaben.

Durch die im vorigen § dargelegten Schwierigkeiten veranlasst, suchen wir nun das in der Unvereinbarkeit der Postulate a), b) und c) des vorigen § bestehende Dilemma dadurch zu lösen, dass wir (c) aufgeben, die durch die Erfahrung gestützten Postulate a) und b) aber beibehalten. Dass das Postulat c) überhaupt aufgegeben werden kann, d. h. dass dasselbe nicht logisch notwendig ist, erkennen wir, indem wir untersuchen, was für eine Bedeutung räumliche und zeitliche Angaben in der Physik haben.

Physikalische Bedeutung räumlicher Angaben. Die Sätze der euklidischen Geometrie bekommen dadurch einen physikalischen Inhalt, dass wir annehmen, es gebe Objekte, welchen die Eigenschaften der Grundgebilde der euklidischen Geometrie zukommen. Wir nehmen an, dass geeignet hergestellte Kanten fester Körper, die äusseren Einwirkungen nicht unterworfen sind, die Definitions-Eigenschaften der Geraden haben (materielle Gerade), und dass der Teil einer materiellen Geraden zwischen zwei ausgezeichneten materiellen Punkten die Eigenschaften der Strecke habe. Die Sätze der Geometrie gehen dann in Sätze über, welche die Gruppierungen materieller Geraden und Strecken betreffen, die bei relativer Ruhe[x] dieser Gebilde möglich sind. Jede materielle Gerade kann vermöge ihrer Eigenschaften fortgesetzt werden. Durch wiederholtes Anlegen einer kürzeren Strecke auf einer längeren und Abzählung dieser Operationen kann letztere durch erstere abgemessen werden, d. h. die Länge der letzteren kann durch eine Zahl ausgedrückt werden, wenn die erstere (Massstab) als ein für allemal gegeben angesehen wird. Die Lage eines materiellen Punktes auf einer materiellen Geraden, auf welcher ein Fundamentalpunkt ein für allemal gegeben ist, kann durch seinen Abstand von letzterem, also durch eine Zahl (Koordinate) festgelegt werden.

[x] Dieser Begriff ist eigentlich unscharf (Rotation). Ich möchte es aber vermeiden, mich bei Dingen aufzuhalten, deren eingehende ~~Untersuchung für das Ziel unserer Über~~ Überlegung für den hier zu verfolgenden Zweck nicht nötig ist.

From among all the material straight lines at rest relative to one another, we think of three that are perpendicular to one another and intersect in one point as being distinguished, and from among all the material straight line segments we likewise imagine a (transportable) segment, which we will think of as being used for measuring all lengths (measuring rod). We call these structures, taken together, "the coordinate system." Geometry teaches how we can express the position of every material point with respect to such a coordinate system *K* by means of three numbers (coordinates). The shape and position of a material structure at rest relative to *K* are given by the totality of the positions (coordinates) of all of its material points with respect to *K*.

The physical meaning of temporal determinations. If we consider what happens physically with respect to the coordinate system *K*, then we know that spatial coordinates *x*, *y*, *z* do not suffice for the determination of the physical variables. The specification of a fourth basic variable, the temporal coordinate, is also needed. This coordinate as well we will define in such a way that each temporal determination shall appear as the result of a precisely defined measurement procedure.

We imagine a completely isolated physical system that repeatedly assumes a specific state *Z*. Then the state *Z* is always followed by the states *Z'*, *Z''*, etc., until the state *Z* is reached again. The system changes its state periodically. We can then count how often the system, which we shall call a "clock," assumes the state *Z*; we will call this number the "temporal determination" of the clock.[47]

We imagine that such a clock is permanently arranged at the point of origin *O* of *K*, and that the spatial dimensions of the clock are so small that, from a geometrical point of view, it may be treated as a "point." By means of the determination of this clock, every event that is spatially infinitely close to the coordinate origin can be assigned a temporal determination, the "time coordinate," or, in brief, the "time" of the event, if we have arbitrarily fixed the zeroth temporal determination of the clock. We cannot evaluate directly the time of an event taking place at another point *A*(*x*, *y*, *z*) of *K* by means of the clock set up at the origin of *K*; for we possess no means, at first, for deciding which temporal determination of the clock set up at the origin of *K* is "simultaneous" with the event, i.e., is to be assigned to it. Therefore, in order to evaluate the events occurring at *A*, we imagine a clock of exactly the same constitution as that at *O* set up at *A*, with the zeroth temporal determination of this clock chosen arbitrarily. By this means we arrive at a temporal valuation of all events occurring at *A*. But with a procedure of this kind there is, as yet, no connection between the temporal determinations concerning the places *O* and *A*; for these temporal determinations to fulfill the demands we impose on temporal determinations in physics, we still have to indicate a procedure by which the clock at *A* can be "regulated according to the one at *O*." To that end we imagine some physical arrangement that makes it possible to send "signals" both from *O* to *A* and from *A* to *O* in such a way that the signal $O - A$ and the signal $A - O$ are to be conceived, for reasons of symmetry, as completely identically constituted processes.

Suppose a signal that was sent from *O* at the determination t_o of the clock at *O* (*O*-time) arrives at *A* at the *A*-time t_A. Let a signal sent from *A* at the *A*-time t_A' arrive at *O* at the *O*-time t_o'. We stipulate that the clock at *A* should be regulated in such a manner that we shall always have

Unter allen relativ zu einander ruhenden ~~geometrischen~~ materiellen Geraden denken wir uns drei auf einander senkrechte und einander in einem Punkt schneidende ausgezeichnet und ebenso unter allen materiellen Strecken eine (transportable), welche wir zur Ausmessung aller Längen benutzt denken (Massstab). Diese Gebilde zusammen nennen wir „Koordinatensystem". Die Geometrie lehrt, wie wir die Lage jedes materiellen Punktes inbezug auf ein derartiges Koordinatensystem K durch drei Zahlen (Koordinaten) ausdrücken können. Die Gestalt und Lage eines materiellen Gebildes, welches relativ zu K ruht, sind gegeben durch die Gesamtheit ~~ist der Inbegriff~~ der Lagen (Koordinaten) aller seiner materiellen Punkte inbezug auf K.

Physikalische Bedeutung zeitlicher Angaben. Betrachten wir das physikalische Geschehen mit Bezug auf das ~~Bezugssyst~~ Koordinatensystem K, so wissen wir dass die physikalischen Veränderlichen durch ~~x y z~~ die Raumkoordinaten x, y, z allein noch nicht bestimmt sind. Es bedarf noch der Angabe einer vierten Grundvariabeln, der ~~Zeitwertes~~ Zeitkoordinate. Auch diese wollen wir so definieren, dass jede Zeitangabe als das Resultat genau definierter Messverfahren erscheint.

Wir denken uns ein nach aussen vollkommen abgeschlossenes physikalisches System, das wiederholt einen bestimmten Zustand Z annimmt. Dann werden auf den Zustand Z jeweilen ~~immer wieder dieselben~~ die Zustände Z' Z'' etc. folgen, bis der Zustand Z wieder erreicht wird. ~~Das System ändert seinen Zustand periodisch~~ Wir können dann abzählen, wie oft das System, das wir „Uhr" nennen wollen, den Zustand Z annimmt; diese Zahl wollen wir die „Zeitangabe" der Uhr nennen.

Eine solche Uhr denken wir uns im Anfangspunkte O von K dauernd angeordnet, wobei wir uns die räumlichen Abmessungen der Uhr so klein denken, dass sie in geometrischer Beziehung als „Punkt" behandelt werden darf. Mittelst der Angaben dieser Uhr lässt sich jedem Ereignisse, ~~bezw. Zustande~~ welches räumlich dem Koordinatenursprung unendlich nahe ist eine Zeitangabe zuordnen, die Zeitkoordinate t° oder kurz „Zeit" des Ereignisses, wenn wir die nullte Zeitangabe der Uhr willkürlich festgelegt haben. Ein in einem anderen Punkte A(x, y, z) von K stattfindendes Ereignis können wir mit der im Anfangspunkte von K befindlichen Uhr nicht unmittelbar zeitlich werten, denn wir besitzen zunächst kein Mittel, um zu entscheiden, welche Zeitangabe der im Anfangspunkte von K befindlichen Uhr mit dem Ereignisse „gleichzeitig", d. h. zuzuordnen ist. Zur Wertung der in A stattfindenden Ereignisse denken wir uns deshalb in A eine Uhr von genau der gleichen Beschaffenheit angeordnet wie in O, und die nullte Zeitangabe dieser Uhr willkürlich gewählt. Dadurch erlangen wir eine zeitliche Wertung aller in A stattfindenden Ereignisse. Die die Orte O und A betreffenden Zeitangaben stehen aber bei dieser Art des Vorgehens noch in keinem Zusammenhange miteinander; ~~man erhielte vielmehr~~ damit diese Zeitangaben den Anforderungen gerecht werden, welche wir an physikalische Zeitangaben stellen, müssen wir noch ein Verfahren angeben, nach welchem wir die Uhr in A nach derjenigen in O „richten" können. Zu diesem Zweck denken wir uns ~~zwischen O und A~~ irgend welche physikalische Anordnung, welche gestattet, von O nach A, sowie von A nach O „Signale" senden zu können, derart, dass aus Symmetriegründen das Signal O–A und das Signal A–O als völlig gleichbeschaffene Vorgänge aufzufassen sind.

Ein von O ~~bei~~ bei der Angabe t_0 der Uhr in O (O-Zeit) abgesandtes Signal gelange bei der A-Zeit t_A nach A. Ein von A bei der A-Zeit t'_A abgesandtes Signal gelange bei der O-Zeit t'_0 ~~[illegible]~~ nach O. Wir ~~stellen die Vorschrift~~ schreiben vor, die Uhr in A sei so zu richten, dass stets

$$t_A - t_O = t_O' - t_A'.$$

If this is possible and is accomplished, then we say that the clock at A is synchronized with the clock at O.

Using the same procedure, clocks arranged at rest at other points B, C, etc., and of identical constitution as the above-mentioned clocks, can also be synchronized with the clock at O. If, upon the completion of this procedure, two other arbitrary clocks, e.g., those at points A and B, are synchronized in accordance with the same definition,* then it is finally no longer of consequence that we privileged the clock at O by regulating all of the other clocks according to it. Then the result of synchronizing all the clocks can be characterized as follows: each clock is synchronized with each of the others.

We define the time coordinate of an event taking place at an arbitrary point of K (point event) as the simultaneous reading of the clock set up at this point and regulated according to the given procedure. Two point events (occurring at different points) are simultaneous if their time coordinates are equal.

It is of importance that the determination of a time coordinate, and thereby also the determination of simultaneity, has meaning only if the coordinate system K to which that determination refers is indicated. Because for the measurement of a time coordinate, or the verification of simultaneity, we need a system of identically constructed clocks *that are set up at rest relative to K.* We will designate the coordinate system K, together with the measuring rod and the regulated system of clocks at rest relative to K that belongs to it, as the "reference system Σ."

Relativity of spatial and temporal determinations. We have now completely defined the physical meaning of spatial and temporal determinations with respect to the reference system Σ. Let us now introduce a second reference system, Σ', which is in uniform translational motion with respect to the first one. Spatial and temporal determinations with regard to this second reference system Σ' can be interpreted in exactly the same way as with regard to the original system Σ, and we will stipulate that the measuring rod and the clocks of Σ' are constituted in the same way as the measuring rod and clocks of Σ when these objects are compared with one another in a state of relative rest.

The space-time coordinates of a specific event with respect to $\Sigma(x, y, z, t)$, and those of the same event with respect to $\Sigma'(x', y', z', t')$ are totally independent of one another by the terms of their definition, even if the state of motion of Σ' is given relative to Σ. Hence it does not follow from these definitions that two point events that are simultaneous with respect to Σ must also be simultaneous with respect to Σ'.

Let us consider, further, a body in arbitrary motion with respect to Σ. Obviously, its shape and position with respect to Σ is determined by the totality of the spatial coordinates of the material points of the body at a specific time t (of Σ). A

* This must be the case, as Laue has shown, when signals propagating in opposite directions are equivalent. In the case of two oppositely directed, closed signal chains we should have (using a notation easy to understand): $(t_A - t_O) + (t_B - t_A) + (t_O - t_B) = (t_B' - t_O') + (t_A' - t_B') + (t_O' - t_A')$. But since $t_A - t_O = t_O' - t_A$ and $t_B' - t_O' = t_O - t_B$, therefore also $t_B - t_A = t_A' - t_B'$. [48]

$$t_A - t_O = t_O' - t_A'$$

sei. Ist dies möglich und durchgeführt, dann sagen wir, die Uhr in A sei
mit der Uhr in O gleich gerichtet.

~~Nach demselben~~ Verfahren können wir auch in anderen Punkten (ruhend angeordnete) gleich der bisher genannten beschaffene
B, C etc. ~~befindliche~~ Uhren mit der in O befindlichen Uhr gleich richten.
Sind nach Beendigung dieses Verfahrens auch zwei beliebige andere Uhren, z. B. die
in den Punkten A und B befindlichen nach derselben Definition gleich
gerichtet[x], so spielt es schliesslich keine Rolle mehr, dass wir die in
O befindliche Uhr dadurch bevorzugt haben, dass wir alle anderen nach
ihr gerichtet haben. Wir können dann ~~das~~ das Resultat des Richtens
aller Uhren so charakterisieren: ~~jede~~ Uhr~~en~~ ~~sind~~ ist mit jeder anderen
gleich gerichtet.

Die Zeitkoordinate eines in einem beliebigen Punkte von K stattfindenden Ereignisses (Punktereignis) definieren wir als die gleichzeitige
Angabe der in jenem Punkte angeordneten, nach der gegebenen Vorschrift
gerichteten Uhr. Zwei (in verschiedenen Punkten stattfindende) Punktereignisse
sind gleichzeitig, wenn ihre Zeitkoordinaten gleich sind.

~~Damit ist die physikalische Bedeutung zeitlicher Angaben voll~~
~~kommen festgelegt.~~ ~~Wesentlich~~ Von Wichtigkeit ist, dass die Angabe einer Zeitkoordinate
und damit auch die Angabe der Gleichzeitigkeit nur dann einen Sinn
hat, wenn das Koordinatensystem K angegeben wird, auf welches sich jene
Angabe bezieht. Denn wir bedürfen zur Messung der Zeitkoordinate
bezw. zur Konstatierung der Gleichzeitigkeit eines ~~Uhren~~ Systems von
gleich beschaffenen Uhren, welche relativ zu K ruhend angeordnet
sind. Wir wollen das Koordinatensystem K samt Massstab und zugehörigen
gerichteten, relativ zu K ruhenden Uhrensystem ~~K~~ als das „Bezugssystem Σ" bezeichnen.
Relativität räumlicher und zeitlicher Angaben.

Mit Bezug auf das Bezugssystem Σ ist nun die physikalische
Bedeutung räumlicher und zeitlicher Angaben vollkommen festgelegt.
Man denke sich nun noch ein zweites Bezugssystem Σ' eingeführt, welches
sich bezüglich des ersten in gleichförmiger Translationsbewegung
befindet. Mit Bezug auf das zweite Bezugssystem Σ' lassen räumliche
und zeitliche Angaben genau so deuten wie mit Bezug auf das ursprüngliche System Σ, ~~Jedem Punktereignis kommen mit Bezug auf Σ best~~
und wir wollen festsetzen, dass ~~die~~ die Uhren bezw. der Massstab bezw. die Uhren von Σ' ~~Σ zur Koordinaten bezw. Zeitmessung~~
seien bzw. der Massstab bezw. die ~~genau~~ gleich beschaffen ~~Massstäbe bezw.~~ Uhren ~~benutzt werden sollen~~
~~wie~~ von Σ, falls diese Gegenstände im Zustande relativer Ruhe miteinander
verglichen werden.
~~Relativität geometrischer und zeitlicher Ta~~

Die Raumzeitkoordinaten eines bestimmten Ereignisses bezüglich Σ
(x, y, z, t) und diejenigen desselben Ereignisses bezüglich Σ' (x', y', z', t') sind
ihrer Definition nach vollkommen unabhängig voneinander, auch
dann wenn der Bewegungszustand von Σ' relativ zu Σ ~~vollkommen~~
~~bekannt~~ gegeben ist. Deshalb ~~ist es~~ geht aus jenen Definitionen nicht hervor,
dass zwei Punktereignisse, welche bezüglich Σ gleichzeitig sind, auch
bezüglich Σ' gleichzeitig sein müssten.

Es werde ferner ein inbezug auf Σ beliebig bewegter Körper betrachtet. ~~Unter~~ ~~Sein~~ Die Gestalt und Lage desselben
inbezug auf Σ wird offenbar bestimmt durch die Gesamtheit der räum-
lichen Koordinaten der ~~Körpers~~ materiellen Punkte des Körpers zu einer
bestimmten Zeit t (von Σ). Eine entsprechende Bemerkung gilt bezüglich

---

[x] Dies muss, wie Laue gezeigt hat, der Fall sein, wenn entgegengesetzt verlaufende Signale gleichwertig sind. Denn für zwei entgegengesetzte, geschlossene Signalketten müsste (in leicht verständlicher Bezeichnungsweise) gelten $(t_A - t_0) + (t_B - t_A) + (t_0^x - t_B) = (t_B' - t_0') + (t_A' - t_B') + (t_0'^x - t_A')$. Da aber $t_A - t_0 = t_0'^x - t_A$ und $t_B' - t_0' = t_0^x - t_B$, so ist auch

$$t_B - t_A = t_A' - t_B'$$

corresponding remark can be made about the shape and position of the same body with respect to Σ' at a specific time t' of Σ'. The fact that the definitions of the space-time coordinates with respect to Σ and Σ' are independent of each other entails that no relationships between the shape of the body with respect to Σ and that with respect to Σ' can be given on the basis of these definitions.

If one takes into account these two consequences, one finds oneself unable to derive the transformation equations (II) of §6; one realizes, instead, that these transformation equations are arbitrary. If we replace equations (II) with transformation equations that are consistent not only with the relativity principle but also with the principle of the constancy of the velocity of light, we arrive at the theory presently designated as the "theory of relativity." Only in this way can we reconcile H. A. Lorentz's eminently fruitful extension of Maxwell's theory with the empirically strongly supported relativity principle.*

§9. Derivation of the Lorentz Transformation

The principle of the constancy of the velocity of light demands the existence of a reference system Σ relative to which every light ray propagates in vacuum with velocity c. According to the relativity principle, all reference systems Σ' in uniform translational motion relative to Σ must possess the same property. Together with Laue, we call each such reference system "justified." Now we ask: What kind of transformation equations must obtain between the space-time coordinates x, y, z, t (with respect to Σ) and the space-time coordinates x', y', z', t' (with respect to Σ') of the same point event so that the principle of the constancy of the velocity of light would hold with respect to both systems?

In any case, the functions that are being sought, which express x' etc. as a function of x, y, z, t, must be completely *linear* functions of these variables. We demand this in order to preserve the homogeneity properties of physical space. If one did not make this assumption, then bodies that are at rest, congruent, and identically located with respect to Σ' would be differently shaped or located when referred to Σ; or clocks that are at rest and identically constructed with respect to Σ' would have different or time-dependent rates when referred to Σ. This will become clear

* One could ask oneself whether such specious-looking physical definitions for the time and space coordinates are really necessary, i.e., whether it is really necessary to saddle the beautiful and airy concepts of space and time with unwieldy rigid bodies and clocks. In my opinion it is not *necessary* to proceed in such a way, but it is no doubt *advantageous*. For one can consider x, y, z, t as mere mathematical auxiliary quantities (parameters) that have meaning only insofar as they make the formulation of physical laws easier. The laws formulated with the help of such parameters have then content only insofar as these parameters can be eliminated from several of these laws. If one uses this kind of procedure, one does not change, in principle, anything on the laws if one introduces into them, instead of x, y, z, t, arbitrary functions of these quantities. In fact, our definitions for these coordinates can be looked upon as an advantageous method for the elimination of the latter. To be sure, I believe that the considerations and definitions regarding space and time presented here are adequate only as long as one abstains from fitting gravitation into the system of the theory of relativity; but this opinion shall <be substantiated later> not be substantiated here any further.

der Gestalt und Lage des selben Körpers inbezug auf Σ' zu einer bestimmten Zeit t' von Σ'. Aus der Unabhängigkeit der Definition der Raum-Zeit-Koordinaten bezüglich Σ und Σ' folgt, dass aus diesen Definitionen Beziehungen zwischen der Gestalt des Körpers inbezug auf Σ und derjenigen inbezug auf Σ' nicht angegeben werden können. ~~Berücksichtigt man diese~~

Berücksichtigt man diese beiden Konsequenzen, so sieht man sich ausser Stande, die Transformationsgleichungen (II) des § 6 abzuleiten. ~~Man~~ erkennt vielmehr das diese Transformationsgleichungen willkürlich sind. Indem wir an Stelle der Gleichungen (II) solche Transformationsgleichungen setzen, welche nicht nur mit dem Relativitätsprinzip sondern auch mit dem Prinzip von der Konstanz der Lichtgeschwindigkeit im Einklang sind, gelangen wir zu der gegenwärtig als „Relativitätstheorie" bezeichneten Theorie. Nur auf diesem Wege können wir die eminent fruchtbare H. A. Lorentz'sche Fortsetzung von Maxwells Theorie mit dem durch die Erfahrung[x] mächtig gestützten Relativitätsprinzip in Einklang bringen.

§9.

~~Ableitung~~ der Lorentz-Transformation.

Das Prinzip von der Konstanz der Lichtgeschwindigkeit fordert, dass es ein Bezugssystem Σ gebe, relativ zu dem sich jeder Vakuum-Lichtstrahl mit der Geschwindigkeit c ausbreitet. Nach dem Relativitätsprinzip muss allen Bezugssystemen Σ', die sich relativ zu Σ im Zustande gleichförmiger Translationsbewegung befinden dieselbe Eigenschaft zukommen. Wir nennen mit Laue jedes derartige Bezugssystem ein „berechtigtes". Wir fragen nun: Was für Transformationsgleichungen müssen zwischen den Raum-Zeit-Koordinaten x, y, z, t (bezüglich Σ) und den ~~der~~ Raum-Zeit-Koordinaten x', y', z', t' (bezüglich Σ') desselben Punktereignisses bestehen, damit ~~in~~ bezüglich beider Systeme das Prinzip der Konstanz der Lichtgeschwindigkeit gelte?

Die gesuchten ~~Gleichungen~~ Funktionen, welche x' etc. in Funktion von x, y, z, t ausdrücken müssen jedenfalls in diesen Variabeln (ganze) lineare (Funktionen) sein. Dies verlangen wir, um die Homogenitätseigenschaft des physikalischen Raumes zu wahren. Würde man diese Annahme nicht machen, so würden bezüglich Σ' ruhende kongruente und ~~bezüglich~~ ~~der Bewegungsrichtung gleich orientierte Körper~~ gleichliegende Körper verschieden gestaltet bezw. gelagert sein, insofern sie auf Σ bezogen werden, ~~und~~ oder würden bezüglich Σ' ruhende, gleich beschaffene Uhren, verschiedene, bezw. von der Zeit abhängige Gang-Geschwindigkeiten besitzen, insofern sie auf Σ bezogen werden.

---

[x] Man könnte sich fragen, ob derartige spitzfindig ~~aussehende~~ physikalische Definitionen für die Zeit- und Raumkoordinaten wirklich nötig sind, d. h. ob es wirklich nötig ist, die schönen und luftigen Begriffe von Raum und Zeit mit schwerfälligen starren Körpern und Uhren zu belasten. Nach meiner Meinung ist es nicht nötig, wohl aber vorteilhaft, so vorzugehen. Man kann nämlich x, y, z, t auch als blosse mathematische Hilfsgrössen (Parameter) betrachten ~~die nur~~ insofern eine Bedeutung haben, als sie ~~in allen physikalischen Relationen eine Rolle spielen auftret~~ die Formulierung physikalischer Gesetze erleichtern. Die mit Hilfe solcher Parameter formulierten Gesetze haben dann nur insofern ~~Bedeutung~~ einen Inhalt, als diese Parameter aus mehreren von ihnen eliminiert werden können. ~~Es~~ Man ändert bei dieser Behandlungsweise prinzipiell nichts an den Gesetzen, wenn man in dieselben statt x, y, z, t beliebige Funktionen dieser Grössen einführt. Unsere Definitionen für diese Koordinaten ~~ist~~ kann man tatsächlich als eine ~~in allen Fällen brauchbare~~ vorteilhafte Methode ~~ansehen~~ zur Elimination derselben ansehen. Allerdings glaube ich, dass die hier gegebenen Überlegungen und Definitionen über Raum und Zeit ~~nur so lange~~ ausreichen, als man auf die Einfügung der Gravitation in das System der Relativitätstheorie verzichtet. Diese Meinung soll ~~später begründet werden~~ aber hier nicht weiter begründet werden.

later on, when we discuss the physical meaning of the Lorentz transformation. We can further demand, without reducing their generality, that the transformation equations be homogeneous, because all that is needed for this is that the path described by the origin of Σ' with respect to Σ pass through the origin of Σ, and that the origin of the time scales in Σ and Σ' be chosen in such a way that the clocks located at the origins of the systems Σ and Σ' both read zero at the moment when the two points coincide.

Suppose that at this moment of the coinicidence of the two origins a vacuum light signal is sent from O or O', which, according to the principle of the constancy of the velocity of light, propagates in a spherical wave with respect to both systems then the spatial points that are just reached by the signal at times t and t' with respect to Σ and Σ', respectively, will be determined by the equations

$$\sqrt{x^2 + y^2 + z^2} = ct$$

$$\text{and } \sqrt{x'^2 + y'^2 + z'^2} = ct'.$$

This means that the equations

$$x^2 + y^2 + z^2 - c^2t^2 = 0$$

$$\text{and } x'^2 + y'^2 + z'^2 - c^2t'^2 = 0$$

must be equivalent. Thus, the transformation equations that we are seeking must be so constituted that the second equation turns into the first one if x', y', z', t' are replaced by their expressions in terms of x, y, z, t. The transformation must therefore make the equation

$$\lambda^2(x^2 + y^2 + z^2 - c^2t^2) = (x'^2 + y'^2 + z'^2 - c^2t'^2) \qquad \ldots (15)$$

into an identity, where all that we know about the factor $\lambda 2$ for the time being is that it must not vanish. But one can see that $\lambda 2$ must be independent of x, y, z, t, for otherwise the right-hand side divided by $\lambda 2$ could not be a homogeneous, complete function of second order in x, y, z, t after the substitution is carried out. For now we will examine the substitution for the case where $\lambda 2 = 1$ and we will show later that from a physical point of view this is the only case deserving of consideration. Instead of (15), we then have

$$x^2 + y^2 + z^2 - c^2t^2 = x'^2 + y'^2 + z'^2 - c^2t'^2 \qquad \ldots (15a)$$

If one introduces the variable $u = ict$ or $u' = ict'$ in place of the time variables t, where i denotes the imaginary unit, one obtains, instead of (15a), the form

$$x^2 + y^2 + z^2 + u^2 = x'^2 + y'^2 + z'^2 + u'^2 \qquad \ldots (15b)$$

As is well known, this choice of time variables derives from Minkowski.[49] Its great significance consists in the fact that by means of it, equation (15a), which governs the substitution that we are seeking, is brought into a form into which the spatial coordinates and the temporal coordinate enter in the same manner.

Let the coefficients of the substitution that we are seeking be denoted as in the accompanying array;[50] the second horizontal row, for example, shall signify that the equation

$$y' = \alpha_{21}x + \alpha_{22}y + \alpha_{23}z + \alpha_{24}u$$

obtains. Obviously, those from among these coefficients that do not contain the index "4" or contain it twice are real, the rest being purely imaginary

	x	y	z	u
x'	α_{11}	α_{12}	α_{13}	α_{14}
y'	α_{21}	α_{22}	α_{23}	α_{24}
z'	α_{31}	α_{32}	α_{33}	α_{34}
u'	α_{41}	α_{42}	α_{43}	α_{44}

Dies wird nachher bei Diskussion der physikalischen Bedeutung der Lorentztransformation klar werden. Wir können ~~es~~ ferner ohne Beschränkung der Allgemeinheit fordern, dass die Transformationsgleichungen homogen seien, denn es ist hiezu nur notwendig, dass die Bahn, die der Anfangspunkt von Σ' bezüglich Σ beschreibt, durch den Anfangspunkt von Σ hindurchgehe, und dass der Anfangspunkt der Zeitskala in Σ und Σ' so gewählt werde, dass die in den Anfangspunkten der Systeme Σ und Σ' befindlichen Uhren im Momente der Koinzidenz beider Punkte auf null zeigen.

Denken wir uns in diesem Augenblick des Zusammenfallens der Anfangspunkte ein Vakuum-Lichtsignal von O bezw. O' ausgesendet, das sich nach dem Prinzip der Konstanz der Lichtgeschwindigkeit ~~bezüglich~~ beider Systeme in einer Kugelwelle ausbreitet, so werden die ~~bezüglich~~ Σ bezw. Σ' vom Signal zur Zeit t bezw. t' erreichten Raumpunkte ~~falls~~ durch die Gleichungen

$$\sqrt{x^2+y^2+z^2} = ct$$

bezw.

$$\sqrt{x'^2+y'^2+z'^2} = ct'$$

bestimmt sein. Es bedeutet dies, dass die Gleichungen

$$x^2+y^2+z^2-c^2t^2 = 0$$

und

$$x'^2+y'^2+z'^2-c^2t'^2 = 0$$

gleichbedeutend sein müssen. Die gesuchten Transformationsgleichungen müssen also so beschaffen sein, dass die zweite Gleichung in die erste übergeht, wenn für x', y', z', t' ihre Ausdrücke in x, y, z, t eingesetzt werden. Die Transformation muss daher die Gleichung

$$\lambda^2(x^2+y^2+z^2-c^2t^2) = (x'^2+y'^2+z'^2-c^2t'^2) \quad \ldots\ldots (15)$$

zu einer Identität machen, wobei wir von dem Faktor λ^2 zunächst nur wissen, dass er nicht verschwinden darf. Man sieht aber, dass λ^2 ~~eine Konstante~~ von x, y, z, t unabhängig sein muss, da sonst die durch λ^2 dividierte rechte Seite nach Ausführung der Substitution nicht eine homogene, ganze Funktion zweiten Grades von x, y, z, t sein könnte. Wir untersuchen nun die Substitution vorläufig für den Fall $\lambda^2 = 1$ und zeigen dann nachträglich, dass dieser Fall ~~der allein~~ vom physikalischen Standpunkt der allein in Betracht kommende ist. Anstelle von (15) tritt dann

$$x^2+y^2+z^2-c^2t^2 = x'^2+y'^2+z'^2-c^2t'^2 \quad \ldots\ldots (15a)$$

Führt man statt der Zeitvariabeln t die Variable $u = ict$ (bezw. $u' = ict'$) ein, wobei i die imaginäre Einheit bezeichnet, so erhält man statt (15a) die Form

$$x^2+y^2+z^2+u^2 = x'^2+y'^2+z'^2+u'^2 \quad \ldots\ldots (15b)$$

Diese ~~Substitution~~ Wahl der Zeitvariablen rührt bekanntlich von Minkowski her. Ihre grosse Bedeutung liegt darin, dass durch sie die die gesuchte Substitution beherrschende Gleichung (15a) auf eine Form gebracht wird, in die die räumlichen Koordinaten und die zeitliche Koordinate in gleicher Weise eingehen.

Die Koeffizienten der gesuchten Substitution seien durch das nebenstehende Schema bezeichnet; es bedeute z. B. die zweite Horizontalreihe, dass die Gleichung

$$y' = \alpha_{21}x + \alpha_{22}y + \alpha_{23}z + \alpha_{24}u$$

bestehen soll. Von diesen Koeffizienten sind offenbar diejenigen, welche den Index „4" nicht oder zweimal enthalten reell, die übrigen rein imaginär.

	x	y	z	u
x'	α_{11}	α_{12}	α_{13}	α_{14}
y'	α_{21}	α_{22}	α_{23}	α_{24}
z'	α_{31}	α_{32}	α_{33}	α_{34}
u'	α_{41}	α_{42}	α_{43}	α_{44}

Equation (15b) is satisfied identically when the following relations exist between the coefficients α:

$$\left.\begin{aligned} \alpha_{1\nu}^2 + \alpha_{2\nu}^2 + \alpha_{3\nu}^2 + \alpha_{4\nu}^2 &= 1 \quad (\nu = 1, 2, 3 \text{ or } 4) \\ \alpha_{1\mu}\alpha_{1\nu} + \alpha_{2\mu}\alpha_{2\nu} + \alpha_{3\mu}\alpha_{3\nu} + \alpha_{4\mu}\alpha_{4\nu} &= 0 \quad (\mu \neq \nu;\ \text{both } \mu \text{ and } \nu \text{ one of the numbers 1-4}) \end{aligned}\right\} \quad (16)$$

This is a total of $4 + \frac{4 \cdot 3}{2} = 10$ conditions that the coefficients have to satisfy. Hence the substitution contains only 6 mutually independent determinations. This must also be so; because three determinations are needed to determine the orientation of Σ' with respect to Σ, and three more are needed for the determination of the magnitude and direction of its velocity.

If one forms $\alpha_{11}x' + \alpha_{21}y' + \alpha_{31}z' + \alpha_{41}u'$ and replaces x', y', z', u' by their expressions in terms of x, y, z, u, then one obtains x as the result. The situation is analogous with the other vertical rows of the above array. Thus, the array also yields the inverse substitution, which expresses x etc. by means of x', y', z', u'. Hence, the quantities α must also satisfy those conditional equations that are analogous to equations (16) in that merely the vertical and the horizontal rows change their roles.

As we already can see from equation (15b) which determines them, the transformations we seek are exactly the same as those we have to apply to the spatial coordinates when passing from an orthogonal coordinate system to another one with the same origin, the only difference being that here one deals with a four-dimensional manifold rather than with a three-dimensional manifold as in the other case. This knowledge forms the basis of Minkowski's four-dimensional treatment of the theory of relativity, which brought about a splendid simplification of the system of the theory of relativity.[51] We shall go into this in greater detail in the next chapter, while in this chapter we will derive the most important results of the theory of relativity in the most elementary way, in order for its physical relationships to emerge most clearly.

The simplest Lorentz transformations are those in which, in addition to the time coordinate, only *one* of the spatial coordinates (e.g., the X-coordinate) undergoes a transformation. This transformation has the accompanying array, where there are, by virtue of (16), three mutually independent relations between the still undetermined coefficients α. Hence we can express these coefficients through *one* quantity β, which will be real and whose value will lie between +1 and −1, if we set

	x	y	z	u
x'	α_{11}	0	0	α_{14}
y'	0	1	0	0
z'	0	0	1	0
u'	α_{41}	0	0	α_{44}

$$\alpha_{11} = \frac{1}{\sqrt{1-\beta^2}} \qquad \alpha_{14} = \frac{i\beta}{\sqrt{1-\beta^2}}$$

$$\alpha_{41} = \frac{-i\beta}{\sqrt{1-\beta^2}} \qquad \alpha_{44} = \frac{1}{\sqrt{1-\beta^2}}$$

where the square root is to be taken with a positive sign.
As a consequence of this choice, equations (16) are satisfied. Conversely, equations (16) determine these values for the coefficients α up to the sign of three of them.*
The corresponding transformation equations are

* By disposing of these algebraic signs, we dispose only of the *direction* in which the x' and u' components are to be taken as positive.

Die Gleichung (15b) ist identisch erfüllt, wenn zwischen den Koeffizienten α die Beziehungen bestehen

$$\left.\begin{aligned} \alpha_{1\nu}^2 + \alpha_{2\nu}^2 + \alpha_{3\nu}^2 + \alpha_{4\nu}^2 &= 1 \quad (\nu = 1, 2, 3 \text{ oder } 4) \\ \alpha_{1\mu}\alpha_{1\nu} + \alpha_{2\mu}\alpha_{2\nu} + \alpha_{3\mu}\alpha_{3\nu} + \alpha_{4\mu}\alpha_{4\nu} &= 0 \quad (\mu \neq \nu;\ \text{sowohl } \mu \text{ als } \nu \text{ eine der Zahlen } 1\text{–}4) \end{aligned}\right\} (16)$$

Dies sind im Ganzen $4 + \frac{4 \cdot 3}{2} = 10$ Bedingungen, welchen die Koeffizienten zu genügen haben. Die Substitution ~~ist~~ enthält daher nur 6 voneinander unabhängige Angaben. Dies muss auch sein, denn es bedarf dreier Angaben zur Bestimmung der Orientierung von Σ' gegen Σ und weiterer drei zur Angabe der Grösse und Richtung seiner Geschwindigkeit.

Bildet man $\alpha_{11}x' + \alpha_{21}y' + \alpha_{31}z' + \alpha_{41}u'$, indem man für x', y', z', u' ihre Ausdrücke in x, y, z, u einsetzt, so erhält man x als Resultat. Analog ist es bei den übrigen Vertikalreihen des obigen Schemas. Dieses liefert also auch die inverse Substitution, welche x etc. durch x', y', z', u' ausdrückt. Die Grössen α müssen also auch jene Bedingungsgleichungen erfüllen, welche den Gleichungen (16) analog sind, indem ~~jene~~ nur Vertikal- und Horizontalreihen ihre Rollen wechseln.

Die von uns gesuchten Transformationen sind, wie schon die sie bestimmende Gleichung (15b) erkennen lässt, genau die gleichen wie diejenigen, welche man auf die räumlichen Koordinaten anzuwenden hat, wenn man von einem rechtwinkligen Koordinatensystem zu einem anderen mit dem gleichen Anfangspunkte übergeht, mit dem einzigen Unterschiede, dass man es hier mit einer vierfachen statt wie dort mit einer dreifachen Mannigfaltigkeit zu thun hat. Auf diese Erkenntnis gründet sich Minkowskis vierdimensionale Behandlung der Relativitätstheorie, die eine grossartige Vereinfachung des Systems der Relativitätstheorie mit sich brachte. Wir wollen auf diese im nächsten Kapitel näher eingehen, in diesem Kapitel aber die ~~einfachsten~~ wichtigsten Ergebnisse der Relativitätstheorie auf elementarstem Wege ableiten, damit die physikalischen Zusammenhänge recht deutlich hervortreten.

Die einfachsten ~~der~~ Lorentz-Transformationen sind solche, bei denen ausser der Zeitkoordinate nur eine der Raumkoordinaten (z. B. die x-Koordinate) eine ~~[illegible]~~ Transformation erfährt. Sie hat das ~~Schema~~ nebenstehende Schema, wobei zwischen den noch unbestimmten Koeffizienten α vermöge (16) drei voneinander unabhängige Relationen bestehen. Man kann diese Koeffizienten ~~die reell [illegible]~~ daher durch eine Grösse β, die reell sei und deren Wert zwischen +1 und −1 liegt, ausdrücken, indem man setzt

	x	y	z	u
x'	α_{11}	0	0	α_{14}
y'	0	1	0	0
z'	0	0	1	0
u'	α_{41}	0	0	α_{44}

$$\alpha_{11} = \frac{1}{\sqrt{1-\beta^2}} \qquad \alpha_{14} = \frac{i\beta}{\sqrt{1-\beta^2}}$$

$$\alpha_{41} = \frac{-i\beta}{\sqrt{1-\beta^2}} \qquad \alpha_{44} = \frac{1}{\sqrt{1-\beta^2}}$$

wobei die Wurzel mit dem positiven Vorzeichen zu nehmen ist. Durch diese Wahl sind die Gleichungen (16) erfüllt. Umgekehrt bestimmen die ~~Koeffizienten~~ Gleichungen (16) diese Werte für die Koeffizienten α bis auf die Vorzeichen dreier von ihnen.[x] Die zugehörigen Transformationsgleichungen lauten

[x] Indem wir über diese Vorzeichen verfügen, verfügen wir nur über den Sinn, in dem die x'- und u'-Komponente positiv zu nehmen sind.

$$\left.\begin{aligned} x' &= \frac{x + i\beta u}{\sqrt{1-\beta^2}} \\ y' &= y \\ z' &= z \\ u' &= \frac{u - i\beta x}{\sqrt{1-\beta^2}} \end{aligned}\right\} \qquad \text{(IIa)}$$

or, written with the real time variable t,

$$\begin{aligned} x' &= \frac{x - \beta ct}{\sqrt{1-\beta^2}} \\ y' &= y \\ z' &= z \\ t' &= \frac{t - \frac{\beta}{c}x}{\sqrt{1-\beta^2}} \end{aligned}$$

From these equations one can easily grasp the significance of the constant β. For we have, for the origin of Σ', permanently $x' = x - \beta ct = 0$. Thus, $\beta c = v$ is the magnitude of the velocity with which this origin glides along the X-axis of Σ. If one substitutes this velocity, then the transformation equations assume the form

$$\left.\begin{aligned} x' &= \frac{x - vt}{\sqrt{1-\frac{v^2}{c^2}}} \\ y' &= y \\ z' &= z \\ u' &= \frac{t - \frac{v}{c^2}x}{\sqrt{1-\frac{v^2}{c^2}}} \end{aligned}\right\} \qquad \text{(IIb)}$$

This system is called the "special Lorentz transformation." These are the equations that, according to the theory of relativity must take the place of equations (II) of §6.

If one solves these equations for x, y, z, t, one obtains equations of the same form, except that v is replaced by $(-v)$. Thus, Σ moves with velocity $-v$ relative to Σ'. If one performs several Lorentz transformations in succession, one obtains again a Lorentz transformation; i.e., the entire set of the Lorentz transformations forms a group. One derives this from (15a) without calculation; because for all such transformations the expression $x^2 + y^2 + z^2 - c^2t^2$ is an invariant.

We can now also formulate the relativity principle in the following way. *The theory of relativity requires that systems of equations in physics turn into systems of equations of the same form if one transforms them by means of the Lorentz transformation.*

Simple calculation shows that the fundamental equations of classical mechanics (§6) do not have this property. Thus, they are not compatible with the theory of relativity.

$$\left.\begin{aligned} x' &= \frac{x + i\beta u}{\sqrt{1-\beta^2}} \\ y' &= y \\ z' &= z \\ u' &= \frac{u - i\beta x}{\sqrt{1-\beta^2}} \end{aligned}\right\} \quad (\text{II}\,a)$$

oder mit reeller Zeitvariable t geschrieben

$$x' = \frac{x - \beta c t}{\sqrt{1-\beta^2}}$$

$$y' = y$$

$$z' = z$$

$$t' = \frac{t - \frac{\beta}{c} x}{\sqrt{1-\beta^2}}$$

Diese Gleichungen lassen ~~zunächst~~ leicht die Bedeutung der Konstante β erkennen.
Für den Anfangspunkt von Σ' ist nämlich dauernd $x' = x - \beta c t = 0$. Es ist also $\beta c = v$
die Grösse der Geschwindigkeit, mit der dieser Anfangspunkt längs der
X Achse von Σ gleitet. Führt man diese Geschwindigkeit ein, so nehmen
die Transformationsgleichungen die Form an

$$\left.\begin{aligned} x' &= \frac{x - vt}{\sqrt{1-\frac{v^2}{c^2}}} \\ y' &= y \\ z' &= z \\ t' &= \frac{t - \frac{v}{c^2} x}{\sqrt{1-\frac{v^2}{c^2}}} \end{aligned}\right\} \quad (\text{II}\,b)$$

Dies System nennt man die „spezielle Lorentz-Transformation". ~~Bew~~
~~wie~~ Es sind dies die Gleichungen, welche nach der Relativitätstheorie an die Stelle der
Gleichungen (II) des § 6 treten müssen. ~~§ 17~~

Löst man diese Gleichungen nach x, y, z, t auf, so erhält man Gleichungen
derselben Gestalt, nur dass v durch $(-v)$ ersetzt ist. Es ist also Σ mit der Geschwindigkeit
$-v$ gegenüber Σ' bewegt. Führt man mehrere Lorentz-Transformationen hinter
einander aus, so erhält man wieder eine Lorentz-Transformation, d. h.
die Gesamtheit der Lorentz-Transformationen bildet eine Gruppe. Dies
folgert man ohne ~~weitere~~ Rechnung aus (15a); denn für alle derartigen
Transformationen ist der Ausdruck $x^2 + y^2 + z^2 - c^2t^2$ eine Invariante.

Wir können nun das Relativitätsprinzip auch so aussprechen. Die
Relativitätstheorie verlangt, dass die Gleichungssysteme der Physik, in Gleichungssysteme von derselben Form übergehen, wenn man sie mittelst
der Lorentz-Transformation transformiert.

Man erkennt durch einfache Rechnung, dass die Grundgleichungen
der klassischen Mechanik (§ 6) diese Eigenschaft nicht haben. Sie sind
also mit der Relativitätstheorie nicht vereinbar.

[The calculations opposite appear on the verso of manuscript page 27. They are similar to ones on page 5 of Einstein's research notes on a generalized theory of relativity (see doc. 10, *Einstein 1995*).]

§10. The Physical Content of the Lorentz Transformation

For x' and t' in (IIb) to be real, it is necessary that $|v|<c$. The translational velocity of Σ' relative to Σ must be smaller than the velocity of light in vacuum. Thus, according to the theory of relativity, it is impossible in principle for a body (coordinate system) to move with superluminal velocity; it will become evident later that from the standpoint of dynamics this means that the kinetic energy of a body grows to infinity when its velocity approaches c.

If a body moves with Σ', i.e., is at rest relative to Σ', what is its shape with respect to Σ? Suppose that the body is a sphere of radius R that is at rest with respect to Σ' and that the equation of its surface is

$$x'^2 + y'^2 + z'^2 = R^2.$$

If we introduce the variables x, y, z, t into this equation by means of (IIb), we obtain the equations of the (moving) surface of the body with respect to Σ. If, in this equation, we set t = const., e.g., $t = 0$, we obtain for the equation of the surface of the body at time $t = 0$ of Σ

$$\frac{x^2}{\left(R\sqrt{1-\frac{v^2}{c^2}}\right)^2} + \frac{y^2}{R^2} + \frac{z^2}{R^2} = 1.$$

$$dx^2+dy^2+dz^2-c^2dt^2 = dx'^2+dy'^2+dz'^2-c^2dt'^2$$

$$dx' = \alpha_{11}\,dx + \alpha_{12}\,dy + \alpha_{13}\,dz + \alpha_{14}\,dt$$
$$dy' = - - - - - -$$
$$dz' = - - - - - -$$
$$dt' = - - - - - -$$

$$\alpha_{11}^2 + \alpha_{41}^2 = 1$$
$$\alpha_{14}^2 + \alpha_{44}^2 = 1$$
$$\alpha_{11}\alpha_{14} + \alpha_{41}\alpha_{44} = 0$$

$$x' = \frac{x + i\beta ct}{\sqrt{1-\beta^2}} \qquad \alpha_{11} = \frac{1}{\sqrt{1-\beta^2}} \qquad \alpha_{14} = \frac{+i\beta}{\sqrt{\ }}$$
$$ct' = \frac{ct \mp i\beta x}{\sqrt{1-\beta^2}} \qquad \alpha_{41} = \frac{-i\beta}{\sqrt{\ }} \qquad \alpha_{44} = \frac{1}{\sqrt{\ }}$$

$$dx_1^2 + \dots + dx_4^2 = A^2(dx_1'^2 + \dots dx_4'^2)$$

$$\begin{matrix} \alpha_{11} & \cdots & \alpha_{14} \\ \alpha_{21} & \cdots & \alpha_{24} \\ - & - & - \\ \alpha_{41} & \cdots & \alpha_{44} \end{matrix}$$

$$\frac{\partial \varphi_1}{\partial y} \qquad \left(\frac{\partial \varphi_1}{\partial x_1}\right)^2 + \cdot + \cdot + \cdot = 1$$

$$\alpha_{11}\,(\alpha_{11}x + \alpha_{12}y + x$$
$$\alpha_{21}$$
$$\alpha_{31}$$
$$\alpha_{41}$$

$$\frac{\partial^2 f}{\partial x^2} + \frac{\partial^2 f}{\partial y^2} = 0$$
$$\frac{\partial f}{\partial x} = p \qquad \frac{\partial f}{\partial y} = q \qquad \frac{\partial p}{\partial x} + \frac{\partial q}{\partial y} = 0$$

$$\frac{\partial^2 f}{\partial x \partial y} = a \qquad \frac{\partial^2 f}{\partial x^2} = b$$
$$\frac{\partial f}{\partial x} = bx + \varphi(y)$$
$$f = b\frac{x^2}{2} + x\varphi(y) + \psi(y)$$
$$\varphi'(y) = a \qquad \varphi(y) = ay + b$$

$$\frac{\partial \varphi}{\partial x} = \omega_1(x, y) \qquad \frac{\partial \varphi}{\partial y} = \omega_2(x, y)$$

geht ~~im Allgemeinen~~ nicht.

§10 Physikalischer Inhalt der Lorentz-Transformation. (28)

Damit x' und t' in (II b) reell seien, ist es notwendig, dass $|v| < c$ sei. Die Translationsgeschwindigkeit von Σ' relativ zu Σ muss kleiner sein als die Vakuum-Lichtgeschwindigkeit. Es ist also nach der Relativitätstheorie prinzipiell ausgeschlossen, dass ein Körper (Koordinatensystem) mit „Überlichtgeschwindigkeit" bewegt wird. Es wird sich später zeigen, dass dies vom dynamischen Standpunkt aus so zu verstehen ist, dass die kinetische Energie eines Körpers bei Annäherung seiner Geschwindigkeit an c ins Unendliche wächst.

Welches ist die auf Σ bezogene Gestalt eines Körpers, der sich mit Σ' bewegt, d. h. relativ zu Σ' ruht? Der Körper sei eine in bezug auf Σ' ruhende Kugel vom Radius R mit der Oberflächengleichung

$$x'^2 + y'^2 + z'^2 = R^2.$$

Führen wir in diese Gleichung die Variabeln x, y, z, t mittelst (II b) ein, so erhalten wir die Gleichungen der (bewegten) Oberfläche des Körpers in bezug auf Σ. Setzen wir in dieser Gleichung $t = \text{konst.}$, z. B. $t = 0$, so erhalten wir als Gleichung der Körperoberfläche zur Zeit $t = 0$ von Σ

$$\frac{x^2}{\left(R\sqrt{1-\frac{v^2}{c^2}}\right)^2} + \frac{y^2}{R^2} + \frac{z^2}{R^2} = 1$$

Thus, with respect to Σ, the body is an ellipsoid of rotation with axes $R\sqrt{1 - \frac{v^2}{c^2}}, R, R$. Thus, by being set into motion, the body changes its shape to some extent, even though only for a reference system that does not take part in the body's motion. This change of shape consists in a $1 : \sqrt{1 - \frac{v^2}{c^2}}$ contraction of all lengths in the direction of the motion. H. A. Lorentz had already introduced such a contraction hypothetically in order to explain the negative result of the Michelson and Morley[52] experiment from the standpoint of his theory.

If V_0 denotes the "real" volume or "rest-volume" of the sphere considered above, i.e., the volume determined by an observer who moves with the sphere (the volume with respect to Σ'), then the volume V with respect to Σ is given by $V = V_0\sqrt{1 - \frac{v^2}{c^2}}$. Thus, in general, the rest-volume V_0 of a body moving with the velocity q with respect to Σ is obtained from its (relative) volume V with respect to Σ according to the equation

$$V_0 = \frac{V}{\sqrt{1 - \frac{q^2}{c^2}}}. \qquad \ldots (17)$$

Let us now consider a clock that is at rest relative to Σ', just like the clock we have introduced there in order to measure time, e.g., the one situated at the coordinate origin. ($x' = y' = z' = 0$) How long does it take with respect to the system Σ (temporal duration Δt) until the temporal determination of the clock under consideration increases by $\Delta t'$? From the fourth of the equations obtainable by reversing (IIb) one gets immediately

$$\Delta t = \frac{\Delta t'}{\sqrt{1 - \frac{v^2}{c^2}}}.$$

Thus, Δt is always greater than $\Delta t'$; i.e., measured with the system of clocks of Σ, the clock moving with velocity v runs slower than it would run if it were at rest (for in the latter case we would have $\Delta t = \Delta t'$). But this astounding consequence holds not only for clocks but also for the lapse of time of arbitrary processes. If we imagine any system that is arranged at rest with respect to Σ and that passes during $\Delta t'$ from an initial state into a final state, then, according to the relativity principle, $\Delta t'$ is independent of the state of motion of Σ', i.e., it is a constant that is characteristic of the process. The equation given above must be valid here as well. Thus, the larger v is, the longer the duration of the process with respect to Σ.

For now, this surprising law of the slowing of temporal changes of arbitrary systems as a consequence of motion has been proved only for rectilinear uniform motions. We cannot know how temporal changes proceed in accelerated systems. But if we consider motions in which the system is accelerated during only a vanishing

Der Körper ist also bezüglich Σ ein Rotationsellipsoid mit den Achsen $R\sqrt{1-\frac{v^2}{c^2}}$,
R, R. Der Körper ändert also dadurch, dass er ~~bew~~ in Bewegung gesetzt
wird, gewissermassen seine Gestalt, allerdings nur für ein Bezugssystem, das
an der Bewegung des Körpers nicht teilnimmt. Diese Gestaltänderung
besteht in ~~einer~~ Verkürzung ~~in der~~ aller Längen in der Bewegungsrichtung
~~im~~ wie $1 : \sqrt{1-\frac{v^2}{c^2}}$. Eine solche Verkürzung hatte H. A Lorentz bereits hypothetisch eingeführt, um vom Standpunkte seiner Theorie das negative
Ergebnis des [Versuchs von] Michelson und Morley zu erklären.

Nennen wir [V_0] das „wirkliche" ~~Volumen~~ [oder „Ruh-Volumen"] der vorhin betrachteten Kugel, d.
h. dasjenige Volumen, welches ein mit der Kugel ~~relativ~~ bewegter Beobachter findet
(Volumen inbezug auf Σ'), so ist das Volumen [V] der Kugel inbezug auf Σ gegeben
durch $V = V_0\sqrt{1-\frac{v^2}{c^2}}$. Man erhält also allgemein das ~~wirkliche~~ [Ruh-] [(relative)] Volumen V_0
eines [inbezug auf Σ] ~~mit~~ der Geschwindigkeit q bewegten Körpers aus dessen Volumen V inbezug
auf Σ nach der Gleichung

$$V_0 = \frac{V}{\sqrt{1-\frac{q^2}{c^2}}} \quad \ldots (12)$$

Betrachten wir ferner eine relativ zu Σ' ruhende Uhr, wie wir sie dort
für die Zeitmessung eingeführt haben, z. B. die im Koordinatenursprung befindliche [$(x'=y'=z'=0)$].
~~Zu welchen Zeiten bezüglich Σ finden~~
~~die [abgelaufenen] („Schläge") der Uhr ist $t' = 0, 1, 2 \ldots$~~
~~die „Schläge" dieser Uhr statt? Aus der ersten und vierten der Gleichungen (II b)~~
~~oder aus der ersten [vierten] Gleichung des zweiten Gleichungssystems finden wir~~

$$\text{~~}t'\text{~~} \quad t = \frac{1}{\sqrt{1-\frac{v^2}{c^2}}} \cdot 1, 2, 3 \ldots$$

~~Ein mit Σ' bewegter, relativ zu der betrachteten Uhr ruhender Beobachter~~
Wie lange dauert es mit Bezug auf das System Σ [(Zeitdauer Δt)] ~~(Δt)~~, bis die Zeitangabe
der betrachteten Uhr um $\Delta t'$ gewachsen ist? Aus der vierten der durch
Umkehrung von (II b) zu gewinnenden Gleichungen folgt ~~für x'=y'=z'=0~~
unmittelbar

$$\Delta t = \frac{\Delta t'}{\sqrt{1-\frac{v^2}{c^2}}}.$$

Es ist also Δt stets grösser als $\Delta t'$, d. h. mit dem Uhrensystem von Σ gemessen
geht die mit der Geschwindigkeit v bewegte Uhr ~~von Σ'~~ langsamer, als
dieselbe Uhr gehen würde, wenn sie unbewegt wäre (denn in letzterem
Fall wäre $\Delta t = \Delta t'$). Diese frappante Konsequenz gilt aber nicht nur für Uhren,
sondern für ~~beliebige zeitliche Verläufe~~ den zeitlichen Ablauf beliebiger
Vorgänge. Denken wir uns irgend ein bezüglich Σ' ruhend angeordnetes
~~(beliebiges) (physikalisches)~~ System, das ~~stets~~ während $\Delta t'$ von einem Anfangszustand in einen Endzustand
übergeht, so ist nach dem Relativitäts[prinzip] ~~theorie~~ $\Delta t'$ unabhängig vom Bewegungszustande von Σ', d. h. eine für den Vorgang charakteristische Konstante. Obige Gleichung
muss auch hier gelten. Es dauert der Vorgang also bezüglich Σ umso länger,
je grösser v ist.

Dies [überraschende] Gesetz der Verlangsamung zeitlicher Änderungen [beliebiger Systeme] infolge der Bewegung ist zunächst nur für geradlinig-gleichförmige Bewegungen bewiesen.
Wir können nicht wissen, wie die zeitlichen Aenderungen in beschleunigten
Systemen vor sich gehen. Betrachten wir jedoch solche Bewegungen, ~~in~~ [bei] denen
das ~~System~~ nur in einem verschwindend kleinen Teil der [gesamten] Bewegungszeit
beschleunigt ist, so dürfen wir von ~~dem Einfluss der Zeiten, in denen das~~

small fraction of the total time of motion, then we may disregard these periods of acceleration insofar as the changes experienced by the system during these periods will be negligible relative to the changes experienced by the system during all of the rest of its motion (motion in a polygon). If the movable system describes a closed polygon with the velocity q relative to Σ *once*, then, in this process, it will experience a change that is determined uniquely (independently of v) by $\sum \Delta t'$.* The time $\sum \Delta t$ that passes with respect to Σ during that process is given by

$$\sum \Delta t = \frac{\sum \Delta t'}{\sqrt{1 - \frac{q^2}{c^2}}}.$$

Thus, the quantity $\sum \Delta t'$ that determines the total change is given by

$$\sum \Delta t' = \sum \Delta t \cdot \sqrt{1 - \frac{q^2}{c^2}}.$$

Thus, if we have two identically constituted systems whose initial states are identical and whose temporal changes are determined by an internal process, and if one of these systems as a whole is at rest relative to Σ, while the other one describes a closed polygon *once*, then, after it has ended its orbit, the second system lags behind relative to the first one with respect to the internal process, and all the more so the faster it was moving in its orbit.

This consequence of the theory of relativity appears to many physicists so fantastic that they believe that they must reject the theory of relativity on its account. Unfortunately, a direct experimental test of this consequence is very difficult to carry out because of the magnitude of c. But such a test is, nonetheless, *possible*; for the apparent proper frequency of moving canal ray ions (with velocities on the order of magnitude of 10^8) must be somewhat smaller than the proper frequency of these ions when they are at rest.—[53]

We must now return to the two assumptions that we made in §9 without justifying them. If the coefficients in the Lorentz transformation were also dependent on x, etc., and t, then it would be not only differences of the spatial and temporal coordinates that entered into propositions about moving bodies and clocks, but their values as well. This is out of the question from a physical standpoint.

Further, in (15) we arbitrarily set $\lambda 2 = 1$. Had we not done this, we would have obtained instead of (II b)

$$x' = \lambda(v)\frac{x - vt}{\sqrt{1 - \frac{v^2}{c^2}}}$$

$$y' = \lambda(v)y$$

$$z' = \lambda(v)z$$

$$t' = \lambda(v)\frac{t - \frac{v}{c^2}x}{\sqrt{1 - \frac{v^2}{c^2}}}.$$

For, since λ is independent of x and t while the special Lorentz transformation is defined by v alone, λ can only depend on v alone. If we introduce a third system Σ'' that possesses the velocity $(-v)$ in the X-direction relative to Σ', we obtain the following transformation equations between Σ and Σ'' by repeated application of the equations we have just written down:

$$x'' = \lambda(v)\lambda(-v)x$$

$$y'' = \lambda(v)\lambda(-v)y$$

$$z'' = \lambda(v)\lambda(-v)z$$

$$t'' = \lambda(v)\lambda(-v)t$$

* The $\Delta t'$ are times measured by a clock moving together with the system. Their sum determines the total change undergone by the system.

~~System beschleunigt ist~~ diesen Zeiten der Beschleunigung insofern absehen, als die Änderungen, welche das System in diesen Zeiten erfährt, relativ zu den Änderungen, welche das System während seiner ganzen ~~Zeit~~ übrigen Bewegung erfährt, zu vernachlässigen sein ~~können~~ werden (Bewegung in einem Polygon). Beschreibt das bewegbare System relativ zu Σ ~~ein~~ einmal ein geschlossenes Polygon mit der Geschwindigkeit q, so wird es dabei eine Änderung erfahren, welche durch $\Sigma\Delta t'$ eindeutig (unabhängig von v) bestimmt ist.* Die Zeit $\Sigma\Delta t$, welche unterdessen inbezug auf Σ vergeht, ist gegeben durch

$$\Sigma\Delta t = \frac{\Sigma\Delta t'}{\sqrt{1-\frac{q^2}{c^2}}}$$

~~Die Zeit $\Sigma\Delta t$, die das System zu einer bestimmten Zustandsänderung~~

Es ist also die die Gesamtänderung bestimmende Grösse $\Sigma\Delta t'$ gegeben durch

$$\Sigma\Delta t' = \Sigma\Delta t \cdot \sqrt{1-\frac{q^2}{c^2}}$$

Liegen also zwei gleich beschaffene und in gleichem Anfangszustande befindliche Systeme vor, deren zeitliche Änderung durch einen inneren Vorgang bestimmt wird, und ruht das eine dieser Systeme als ganzes relativ zu Σ, während das andere einmal ein geschlossenes Polygon beschreibt, so ist, nachdem ~~das zweite System~~ seinen Umlauf ~~beendigt hat~~, relativ zum ~~die Zustandsänderung~~, das zweite System ~~bezüglich~~ ~~des~~ ersten bezüglich des inneren Vorganges zurückgeblieben, und zwar desto mehr, je rascher es auf seinem Umlaufe bewegt war.

Diese Konsequenz der Relativitätstheorie erscheint manchen Physikern so abenteuerlich, dass sie um derselben willen die Relativitätstheorie verwerfen zu müssen glauben. Leider ist eine direkte experimentelle Prüfung derselben wegen der Grösse von c sehr schwer durchzuführen. ~~Die~~ Die Möglichkeit einer solchen Prüfung liegt immerhin vor; es sollte nämlich die scheinbare Eigenfrequenz der ~~bewegten~~ (mit Geschwindigkeiten von der Grössenordnung 10^8) bewegten Kanalstrahlionen etwas kleiner sein als die Eigenfrequenz dieser Ionen, wenn sie ~~sich im Zustand der Ruhe befinden~~ unbewegt sind. —

Wir ~~sind nun in der~~ müssen nun auf die beiden Annahmen zurückkommen, die wir im § 9 gemacht haben, ohne sie zu rechtfertigen. Wären die Koeffizienten der Lorentztransformation noch von x etc. bezw. t abhängig, so würden in die Sätze über bewegte Körper und Uhren ~~neben Koordi~~ nicht nur Differenzen räumlicher und zeitlicher Koordinaten eingehen, sondern auch deren ~~absolute~~ Werte. Dies ist vom physikalischen Standpunkte ausgeschlossen.

Wir haben ferner in (15) willkürlich ~~den Faktor l^2 in (15) gleich~~ $l^2 = 1$ gesetzt. Hätten wir dies nicht gethan, so hätten wir anstelle von (II b) erhalten

$$x' = l(v)\frac{x - vt}{\sqrt{1-\frac{v^2}{c^2}}}$$

$$y' = l(v)\, y$$

$$z' = l(v)\, z$$

$$t' = l(v)\frac{t - \frac{v}{c^2}x}{\sqrt{1-\frac{v^2}{c^2}}}$$

Da l von x und t unabhängig ist, die spezielle Lorentz-Transformation aber durch v allein definiert ist, so kann nämlich l nur von v allein abhängen. Führen wir ein drittes System Σ'' ein, das relativ zu Σ' die Geschwindigkeit $(-v)$ in der X-Richtung besitzt, so erhält man durch nochmalige Anwendung der soeben hingeschriebenen Gleichungen zwischen Σ und Σ'' die Transformationsgleichungen:

$$x'' = l(v)\, l(-v)\, x$$

$$y'' = l(v)\, l(-v)\, y$$

$$z'' = l(v)\, l(-v)\, z$$

$$t'' = l(v)\, l(-v)\, t$$

---

* Die $\Delta t'$ sind Zeiten, gemessen an einer mit dem System bewegten Uhr. Ihre Summe bestimmt die Gesamtänderung, welche das System erleidet.

according to the first three of these equations, the coordinate axes of Σ and Σ'' are permanently coincident. Since the coordinates are measured with identical measuring rods in both systems, we must have $x'' = x$, etc. Hence

$$\lambda(v)\lambda(-v) = 1.$$

Furthermore, the second equation of the transformation shows that $\frac{1}{\lambda(v)}$ is the length, with respect to Σ, of a rod of length 1, that moves with Σ' and is oriented parallel to the y'-axis. But from a physical standpoint it is clear that the length of a rod moving perpendicularly to its extension must be independent of the *orientation* of the motion, which means that we must have

$$\lambda(v) = \lambda(-v).$$

From these two equations it follows that $\lambda = \pm 1$. The special case $v = 0$ shows that the choice of the positive sign is the only one that is possible.

§11. Further Kinematical Consequences of the Transformation Equations

Addition theorem of velocities. Suppose that, relative to the system Σ', a point moves uniformly according to the equations

$$x' = \mathfrak{q}'_x t'$$
$$y' = \mathfrak{q}'_y t'$$
$$z' = \mathfrak{q}'_z t'$$

How does the point move with respect to Σ?

If in these equations of motions we substitute x, y, z, t for x', y', z', t' with the help of the special Lorentz transformation (IIb), we obtain the equations

$$x = \mathfrak{q}_x t$$
$$y = \mathfrak{q}_y t$$
$$z = \mathfrak{q}_z t$$

where we set

$$\left.\begin{aligned} \mathfrak{q}_x &= \frac{\mathfrak{q}'_x + v}{1 + \frac{\mathfrak{q}'_x v}{c^2}} \\ \mathfrak{q}_y &= \sqrt{1 - \frac{v^2}{c^2}}\,\frac{\mathfrak{q}'_y}{1 + \frac{\mathfrak{q}'_x v}{c^2}} \\ \mathfrak{q}_z &= \sqrt{1 - \frac{v^2}{c^2}}\,\frac{\mathfrak{q}'_z}{1 + \frac{\mathfrak{q}'_x v}{c^2}} \end{aligned}\right\} \qquad (18)$$

Thus, equations (18) take the place of the law of the parallelogram of velocities; for here we have added the velocity v in the direction of the positive X-axis and the arbitrarily directed velocity $\mathfrak{q}'$. A simple calculation gives us for the value q of the resulting velocity

$$q^2 = \frac{q'^2 + v^2 + 2q'v\cos\vartheta' - \left(\frac{q'v}{c}\sin\vartheta'\right)^2}{\left(1 + \frac{q'v}{c^2}\cos\vartheta'\right)^2}, \qquad \ldots (18a)$$

where we have set

$$\operatorname{tg}\vartheta' = \frac{\sqrt{\mathfrak{q}'^2_y + \mathfrak{q}'^2_z}}{\mathfrak{q}'_x}$$

Since (18a) can be brought into the form

gemäss den ersten drei dieser Gleichungen fallen die Koordinatenachsen von Σ und Σ'' dauernd zusammen. Da in beiden Systemen die Koordinaten durch gleiche ~~bew~~ Massstäbe gemessen werden, so muss also $x'' = x$ etc. sein. Es ist also

$$\lambda(v)\,\lambda(-v) = 1.$$

Die zweite Gleichung der Transformation lässt ferner erkennen, dass $\frac{1}{\lambda(v)}$ die auf Σ bezogene Länge eines Stabes von der Länge 1 ist, der sich mit Σ' bewegt und parallel der y'-Axe orientiert ist. Es ist aber vom physikalischen Standpunkte aus klar, dass die Länge eines senkrecht zu seiner Ausdehnung bewegten Stabes unabhängig sein muss vom Sinne der Bewegung, das heisst es muss sein

$$\lambda(v) = \lambda(-v).$$

Aus diesen beiden Gleichungen folgt $\lambda = \pm 1$. Der Spezialfall $v=0$ lehrt, dass ~~$\lambda = +1$ sein muss.~~ nur die Wahl des positiven Zeichens möglich ist.

§ 11. Weitere kinematische Folgerungen aus den Transformationsgleichungen.

Additionstheorem der Geschwindigkeiten. Relativ zum System Σ' bewege sich ein Punkt gleichförmig gemäss den Gleichungen

$$x' = q_x' t'$$
$$y' = q_y' t'$$
$$z' = q_z' t'.$$

Wie bewegt sich der Punkt inbezug auf Σ?

~~Setzt man~~ Ersetzt man in diesen Bewegungsgleichungen x', y', z', t' mittelst der speziellen Lorentz' Transformation (IIb) durch x, y, z, t, so erhält man die Gleichungen

$$x = q_x t$$
$$y = q_y t$$
$$z = q_z t,$$

wobei gesetzt ist

$$\left.\begin{aligned} q_x &= \frac{q_x' + v}{1 + \frac{q_x' v}{c^2}} \\ q_y &= \sqrt{1-\frac{v^2}{c^2}}\,\frac{q_y'}{1 + \frac{q_x' v}{c^2}} \\ q_z &= \sqrt{1-\frac{v^2}{c^2}}\,\frac{q_z'}{1 + \frac{q_x' v}{c^2}} \end{aligned}\right\} \quad (18)$$

Die Gleichungen (18) treten also an die Stelle des Gesetzes vom Parallelogramm der Geschwindigkeiten, denn wir haben hier die Geschwindigkeit v in Richtung der positiven X-Achse mit der beliebig gerichteten Geschwindigkeit q' zusammengesetzt. Für die Grösse q der resultierenden Geschwindigkeit erhält man durch einfache Rechnung

$$q^2 = \frac{q'^2 + v^2 + 2q'v\cos\vartheta' - \left(\frac{q'v}{c}\sin\vartheta'\right)^2}{\left(1 + \frac{q'v}{c^2}\cos\vartheta'\right)^2} \quad \ldots (18a),$$

wobei

$$\operatorname{tg}\vartheta' = \frac{\sqrt{q_y'^2 + q_z'^2}}{q_x'}$$

gesetzt ist. Da man (18a) in die Form

$$\frac{q^2}{c^2} = \frac{\left(1 + \frac{q'v}{c}\cos\vartheta'\right)^2 - \left(1 - \frac{q'^2}{c^2}\right)\left(1 - \frac{v^2}{c^2}\right)}{\left(1 + \frac{q'v}{c}\cos\vartheta'\right)^2} \qquad \ldots (18a')$$

and we always have $v < c$, then we also always have $q < c$, if we choose $q' < c$. Thus, we can never attain c by the addition of "subluminal velocities." But if $q' = c$, then it will also always be the case that $q = c$, which, after all, already follows from the principle of the constancy of the velocity of light.

We draw yet another interesting consequence from this addition theorem. If the vector $\mathfrak{q}'$ is parallel to the x' axis of Σ', then

$$q = \frac{q' + v}{1 + \frac{q'v}{c^2}}. \qquad \ldots (18a')$$

If we assume that there is a physical effect that propagates from the place of its excitation with a velocity that is greater than c, then such a propagation must also be possible with respect to Σ', in particular along the X' axis of Σ'. Then it is possible to give to the translational velocity v a negative value that is so large, considered as an absolute value (<c), that $1 + \frac{q'v}{c^2}$ becomes negative. Then, according to (18a'), the propagation velocity q of the signal becomes negative with respect to Σ. I.e., there would exist effects at a distance that precede their cause. Since the existence of such effects is quite improbable,* then, according to the theory of relativity, one will have to consider it out of the question that there is a kind of signal (i.e., a propagation usable in principle for telegraphy) whose velocity exceeds c.

Fizeau's experiment. Mr. Laue was the first to point out that the result of Fizeau's experiment,[54] the deduction of which—as we have seen in §7—is wrought with the greatest difficulties if one avoids the theory of relativity, follows in a very simple fashion from equation (18a'). We have only to assume here that the velocity of light in a given medium, when considered relative to that medium, is independent of how the other bodies are moving. If we put V_0 for this velocity and call q the velocity of the medium, then we may immediately apply equation (18a'), if we stipulate that the medium is at rest relative to Σ'. Then we have to substitute in (18a')

V in place of q
V_0 in place of q'
q in place of v

so that we obtain

$$V = \frac{V_0 + q}{1 + \frac{V_0 q}{c^2}},$$

or, to a first degree of approximation, if we take into account that $V_0 = \frac{c}{n}$:

By combining several signal devices of this kind, one could even achieve that the action that precedes its cause occurs at the location of the cause itself.

$$\left(\frac{q}{c}\right)^2 = \frac{\left(1+\frac{q'v}{c}\cos\vartheta'\right)^2 - \left(1-\frac{q'^2}{c^2}\right)\left(1-\frac{v^2}{c^2}\right)}{\left(1+\frac{q'v}{c}\cos\vartheta'\right)^2} \quad \ldots (18a')$$

bringen kann, und stets $v<c$ ist, so ~~erkennt~~ ist auch stets $q<c$, wenn $q'<c$ gewählt wird. Man kann also durch Zusammensetzung von „Unterlichtgeschwindigkeiten" c niemals erreichen. Ist aber $q'=c$, so ist stets auch $q=c$, was ja schon aus dem Prinzip der Konstanz der Lichtgeschwindigkeit folgt.

Wir ziehen noch eine interessante Konsequenz aus diesem Additionstheorem. ~~Ist~~ Ist der Vektor q' der x' Achse von Σ' parallel, so ist

$$q = \frac{q'+v}{1+\frac{q'v}{c^2}} \quad \ldots (18a')$$

Nehmen wir an, es gebe eine physikalische Wirkung, welche sich von dem Orte ihrer Erregung mit einer Geschwindigkeit ausbreitet, die grösser ist als c, so muss eine derartige Ausbreitung auch inbezug auf Σ' möglich sein, speziell auch längs der X' Axe von Σ'. Es ist dann möglich, der Translationsgeschwindigkeit v einen negativen und absolut genommen so grossen Wert ($<c$) zu geben, dass $1+\frac{q'v}{c^2}$ negativ wird. Dann wird nach (18a') die Ausbreitungsgeschwindigkeit q des Signales inbezug auf Σ negativ. D. h. es gäbe Wirkungen in die Ferne, die ihrer Ursache vorangehen. Da eine Existenz derartiger Wirkungen recht unwahrscheinlich ist,[x] so wird man es nach der Relativitätstheorie wohl für ausgeschlossen ansehen müssen, dass es ~~irgend~~ eine Art Signal~~geschwindigkeit~~ (d. h. eine zum Telegraphieren prinzipiell benutzbare Ausbreitung) gebe, deren Geschwindigkeit c übertrifft.

Fizeau's Experiment. Herr Laue hat zuerst darauf aufmerksam gemacht, dass aus Gleichung (18a') in sehr einfacher Weise das Resultat des Fizeauschen Versuches folgt, dessen Ableitung – wie wir im § 2 gesehen haben – bei Vermeidung der Relativitätstheorie die grössten Schwierigkeiten mit sich bringt. Wir haben hier nur anzunehmen, dass die Lichtgeschwindigkeit in einem Medium relativ zu diesem unabhängig davon ist, ~~wie~~ wie die übrigen Körper bewegt sind. Setzen wir diese Geschwindigkeit V_0 ~~und nehmen wir~~ und nennen wir q die Geschwindigkeit des Mediums, so können wir Gleichung (18a') sogleich anwenden, wenn wir festsetzen, dass das Medium relativ zu Σ' ruhe. Wir haben dann in (18a') einzusetzen

V statt q

V_0 statt q'

q statt v,

sodass wir erhalten

$$V = \frac{V_0+q}{1+\frac{V_0 q}{c^2}}$$

oder in erster Annäherung, wenn wir berücksichtigen, dass $V_0 = \frac{c}{n}$ ist:

[x] Durch Kombination mehrerer derartiger Signaleinrichtungen könnte man sogar erzielen, dass die ihrer Ursache vorangehende Wirkung am Orte der Ursache selbst auftritt.

$$V = V_0 + q\left(1 - \frac{1}{n^2}\right),$$

which corresponds to the result of Fizeau's experiment.

Doppler's principle and aberration. According to the undulatory theory of light, if the light source lies at infinity, then the components of the light vector of a monochromatic light wave in vacuum can be represented by expressions that are proportional to the form

$$\sin\omega\left(t - \frac{lx + my + nz}{c}\right).$$

(ω denotes the 2π-fold frequency, and l, m, n denote the direction cosine of the wave normals.) If the above expression refers to Σ, then the corresponding expression that refers to Σ' reads

$$\sin\omega'\left(t' - \frac{l'x' + m'y' + n'z'}{c}\right),$$

where, because of (IIb), we have the following relations between ω, l, m, n, and ω', l', m', n':

$$\omega' = \omega\frac{1 - l\frac{v}{c}}{\sqrt{1 - \frac{v^2}{c^2}}} \qquad \ldots (19)$$

$$\left.\begin{aligned} l' &= \frac{l - \frac{v}{c}}{1 - l\frac{v}{c}} \\ m' &= \frac{m}{(1 - l\frac{v}{c})\sqrt{1 - \frac{v^2}{c^2}}} \\ n' &= \frac{n}{(1 - l\frac{v}{c})\sqrt{1 - \frac{v^2}{c^2}}} \end{aligned}\right\} \qquad \ldots (20)$$

Equation (19) expresses the Doppler principle,* and the first of equations (20) the aberration law. To get the latter in a clearer form, to a first degree of approximation, we set

$$l' = \cos(\varphi + \delta)$$
$$l = \cos\varphi$$

and obtain, by considering only the terms of first order in $\frac{v}{c}$ and δ

$$\delta = \frac{v}{c}\sin\varphi \qquad \ldots (20')$$

---

* The term appearing in the denominator of the right-hand side is closely related to the arguments of §10. According to the theory of relativity, there exists a second-order Doppler effect if $l = 0$, i.e., if the relative velocity of the observer (Σ') with respect to the light source (Σ) is perpendicular to the connecting line light source-observer.

$$V = V_0 + v\left(1 - \frac{1}{n^2}\right),$$

was dem Ergebnis der Fizeau'schen Experimentes entspricht.

Dopplers Prinzip und Aberration. Nach der Undulationstheorie des Lichtes lassen sich die Komponenten ~~eines~~ des Lichtvektors einer monochromatischen Vakuum-Lichtwelle, falls die Lichtquelle im Unendlichen liegt, durch Ausdrücke darstellen, die der Form

$$\sin \omega \left(t - \frac{lx + my + nz}{c}\right)$$

proportional sind. (Mit ω ist die 2π-fache Periodenzahl, mit l, m, n sind die Richtungskosinus der Wellennormalen bezeichnet). Bezieht sich der angegebene Ausdruck auf Σ, so lautet der entsprechende bezüglich Σ'

$$\sin \omega' \left(t' - \frac{l'x' + m'y' + n'z'}{c}\right),$$

wobei wegen (II b) zwischen ω, l, m, n, und ω', l', m', n' die Beziehungen bestehen:

$$\omega' = \omega \frac{1 - l\frac{v}{c}}{\sqrt{1 - \frac{v^2}{c^2}}} \quad \ldots\ldots (19)$$

$$\left.\begin{aligned} l' &= \frac{l - \frac{v}{c}}{1 - l\frac{v}{c}} \\ m' &= \frac{m}{\left(1 - l\frac{v}{c}\right)\sqrt{1 - \frac{v^2}{c^2}}} \\ n' &= \frac{n}{\left(1 - l\frac{v}{c}\right)\sqrt{1 - \frac{v^2}{c^2}}} \end{aligned}\right\} (20)$$

Gleichung 19) drückt das das Doppler'sche Prinzip aus,ˣ die erste der Gleichungen (20) das Aberrationsgesetz. Um letzteres in übersichtlicher Form in erster Annäherung zu haben, setzen wir

$$l' = \cos(\varphi + \delta)$$
$$l = \cos \varphi$$

und erhalten unter Berücksichtigung nur der Glieder erster Ordnung in $\frac{v}{c}$ und δ

$$\delta = \frac{v}{c} \sin \varphi . \quad \ldots\ldots (20')$$

~~§ 12~~

ˣ Das im Nenner der rechten Seite auftretende Glied ~~drückt~~ hängt aufs engste mit den ~~kinematischen~~ Betrachtungen des § 10 zusammen. Es gibt ~~für den~~ nach der Relativitätstheorie einen Dopplereffekt zweiter Ordnung in dem Falle, dass $l = 0$, d. h. dass die Relativgeschwindigkeit des Beobachters (Σ') gegen die Lichtquelle (Σ) senkrecht steht auf der Verbindungslinie Lichtquelle – Beobachter.

§12. Application of the Lorentz Transformation to the Electromagnetic Field Equations

The foundation of Lorentz's electrodynamics of moving bodies accords with the theory of relativity only if equations (I) of §1 transform to equations of the same form when the Lorentz transformation (IIb) is applied to them. We assume that equations (I) hold for the reference system Σ. For the differential operations $\frac{\partial}{\partial x}, \frac{\partial}{\partial y}, \frac{\partial}{\partial z}, \frac{\partial}{\partial t}$ we substitute the following ones, which are equivalent according to (IIb):

$$\left.\begin{aligned} \frac{\partial}{\partial x} &= b\left(\frac{\partial}{\partial x'} - \frac{v}{c^2}\frac{\partial}{\partial t'}\right) \\ \frac{\partial}{\partial y} &= \frac{\partial}{\partial y'} \\ \frac{\partial}{\partial z} &= \frac{\partial}{\partial z'} \\ \frac{\partial}{\partial t} &= b\left(-v\frac{\partial}{\partial x'} + \frac{\partial}{\partial t'}\right), \end{aligned}\right\} \qquad \text{(IIb')}$$

where we have set $b = \frac{1}{\sqrt{1 - \frac{v^2}{c^2}}}$. In this way we obtain the equations [55]

$$\operatorname{curl}\mathfrak{h}' = \frac{1}{c}(\dot{\mathfrak{e}}' + \mathfrak{q}'\rho') \qquad \operatorname{curl}\mathfrak{e}' = -\frac{1}{c}\dot{\mathfrak{h}}'$$
$$\operatorname{div}\mathfrak{e}' = \rho' \qquad \operatorname{div}\mathfrak{h}' = 0,$$

which are in complete accord with system (I) if we set:

$$\left.\begin{aligned} \mathfrak{e}'_x &= \mathfrak{e}_x & \mathfrak{h}'_x &= \mathfrak{h}_x \\ \mathfrak{e}'_y &= b\left(\mathfrak{e}_y - \frac{v}{c}\mathfrak{h}_z\right) & \mathfrak{h}'_y &= b\left(\mathfrak{h}_y + \frac{v}{c}\mathfrak{e}_z\right) \\ \mathfrak{e}'_z &= b\left(\mathfrak{e}_z + \frac{v}{c}\mathfrak{h}_y\right) & \mathfrak{h}'_z &= b\left(\mathfrak{h}_z - \frac{v}{c}\mathfrak{e}_y\right) \end{aligned}\right\} \qquad (21)$$

$$\rho' = b\left(1 - \frac{v\mathfrak{q}_x}{c^2}\right)\rho \qquad \ldots (22)$$

$$\left.\begin{aligned} \mathfrak{q}'_x &= \frac{\mathfrak{q}_x - v}{1 - \frac{\mathfrak{q}_x v}{c^2}} \\ \mathfrak{q}'_y &= \frac{\mathfrak{q}_y}{b\left(1 - \frac{\mathfrak{q}_x v}{c^2}\right)} \\ \mathfrak{q}'_z &= \frac{\mathfrak{q}_z}{b\left(1 - \frac{\mathfrak{q}_x v}{c^2}\right)} \end{aligned}\right\} \qquad (23)$$

From this we see that the fundamental equations of Lorentz's theory fulfill the reqirements of the theory of relativity. Furthermore, equations (21) and (22) show

~~§12. Die Grundgleichungen der Lorentz'schen Elektrodynamik bewegter Körper und die Relativitätstheorie.~~

§12. Anwendung der Lorentz-Transformation auf die elektromagnetischen Feldgleichungen.

Die Grundlage der Lorentz'schen Elektrodynamik bewegter Körper entspricht nur dann der Relativitätstheorie, wenn die Gleichungen (I) des §1 ~~durch die~~ in Gleichungen der nämlichen Form übergehen, wenn man auf sie die Lorentz'transformation (II b) anwendet. Wir nehmen an, dass die Gleichungen (I) für das Bezugssystem Σ gelten. ~~Gemäss (II b)~~ Wir setzen ~~nun~~ für die Differential-Operationen $\frac{\partial}{\partial x}, \frac{\partial}{\partial y}, \frac{\partial}{\partial z}, \frac{\partial}{\partial t}$ die folgenden nach (II b) gleichwertigen

$$\left.\begin{aligned}
\frac{\partial}{\partial x} &= b\left(\frac{\partial}{\partial x'} - \frac{v}{c^2}\frac{\partial}{\partial t'}\right)\\
\frac{\partial}{\partial y} &= \frac{\partial}{\partial y'}\\
\frac{\partial}{\partial z} &= \frac{\partial}{\partial z'}\\
\frac{\partial}{\partial t} &= b\left(-v\frac{\partial}{\partial x'} + \frac{\partial}{\partial t'}\right),
\end{aligned}\right\}\quad (\text{II } b')$$

wobei $b = \frac{1}{\sqrt{1-\frac{v^2}{c^2}}}$ gesetzt ist. Wir erhalten so die dem System (I) völlig entsprechenden Gleichungen

$$\operatorname{curl} \mathfrak{h}' = \frac{1}{c}(\dot{\mathfrak{e}}' + \mathfrak{q}'\varrho') \qquad \operatorname{curl} \mathfrak{e}' = -\frac{1}{c}\dot{\mathfrak{h}}'$$

$$\operatorname{div} \mathfrak{e}' = \varrho' \qquad \operatorname{div} \mathfrak{h}' = 0,$$

wenn wir setzen:

$$\left.\begin{aligned}
\mathfrak{e}_x' &= \mathfrak{e}_x & \mathfrak{h}_x' &= \mathfrak{h}_x\\
\mathfrak{e}_y' &= b\left(\mathfrak{e}_y - \frac{v}{c}\mathfrak{h}_z\right) & \mathfrak{h}_y' &= b\left(\mathfrak{h}_y + \frac{v}{c}\mathfrak{e}_z\right)\\
\mathfrak{e}_z' &= b\left(\mathfrak{e}_z + \frac{v}{c}\mathfrak{h}_y\right) & \mathfrak{h}_z' &= b\left(\mathfrak{h}_z - \frac{v}{c}\mathfrak{e}_y\right)
\end{aligned}\right\}\quad (21)$$

~~$\varrho' = \frac{\partial \mathfrak{e}_x'}{\partial x'} + \frac{\partial \mathfrak{e}_y'}{\partial y'} + \frac{\partial \mathfrak{e}_z'}{\partial z'}$~~

$$\varrho' = b\left(1 - \frac{v\mathfrak{q}_x}{c^2}\right)\varrho \qquad (22)$$

$$\left.\begin{aligned}
\mathfrak{q}_x' &= \frac{\mathfrak{q}_x - v}{1 - \frac{\mathfrak{q}_x v}{c^2}}\\
\mathfrak{q}_y' &= \frac{\mathfrak{q}_y}{b\left(1 - \frac{\mathfrak{q}_x v}{c^2}\right)}\\
\mathfrak{q}_z' &= \frac{\mathfrak{q}_z}{b\left(1 - \frac{\mathfrak{q}_x v}{c^2}\right)}
\end{aligned}\right\}\quad (23).$$

Wir sehen hieraus, dass die Grundgleichungen der Lorentz'schen Theorie den Forderungen ~~in das Schema~~ der Relativitätstheorie entsprechen. Es zeigen ferner die Gleichungen (21) und (22), wie die Komponenten der Feldstärke und der Dichte der Elektrizität bezüglich Σ' zu berechnen sind, wenn sie inbezug auf Σ bekannt sind. Diese Grössen sind bezüglich Σ' genau so definiert

how the components of the field strength and density of electricity with respect to Σ' are to be calculated if they are known with respect to Σ. These quantities have to be conceived of as being defined in exactly the same manner with respect to Σ' as are the corresponding unprimed quantities with respect to Σ. Equations (23) teach us nothing new; they contain the addition theorem for velocities and are identical with the inverse of equations (18).

We would also obtain a transformation of equations (I) into equations of corresponding form if, everywhere on the right-hand-sides of (21) and (22), we were to add one and the same factor, dependent on v in an arbitrary fashion. But it can be shown in a manner quite similar to the way this was shown for the Lorentz transformation at the end of §10, that this factor must be equal to 1.

One can see from equations (21) that the electric as well as the magnetic field do not possess a *separate* existence if the field is considered from all of the justified systems. For example, a purely magnetic field as observed from Σ also possesses—as observed from Σ'—electrical components, as the second and third of equations (21) show. Thus, the electromagnetic field emerges as a unitary physical structure that is determined by six components for every space-time point. The formal properties of this structure shall be described later <by Minkowski's method>.

Equation (22) shows that the electrical *density* is not an invariant of the Lorentz transformation. By contrast, the *quantity* of electricity, i.e., the product $\rho d\tau$, is invariant, i.e., independent of the choice of the reference system. That is to say, if Σ' is chosen such that it is momentarily at rest relative to the volume element under consideration, then $\mathfrak{q}_x' = 0$, and the inverse* of (22) reads

$$\rho = \frac{\rho'}{\sqrt{1 - \frac{v^2}{c^2}}}.$$

But according to equation (17) we have

$$d\tau_0 = d\tau' = \frac{d\tau}{\sqrt{1 - \frac{v^2}{c^2}}}.$$

From the two equations it follows that $\rho' d\tau' = \rho d\tau$, from which the claim follows. One can also deduce this invariance directly from equations (I) and the relativity principle. For it follows from the first two equations (I) that

$$\operatorname{div}(\mathfrak{q}\rho) + \frac{\partial \rho}{\partial t} = 0$$

This equation expresses the indestructibility of the quantity of electricity, for it asserts that the increase in the quantity of electricity of a unit volume per unit time is equal to the quantity of electricity flowing during this time into the volume under consideration (continuity equation). Thus, if we follow an electrical particle (quantity of electricity ε) on its path between the times t_1 and t_2, then we have $\varepsilon_1 = \varepsilon_2$. This holds, to begin with, for a justified reference system K. For a second justified reference system K$'$ we have, analogously, $\varepsilon'_1 = \varepsilon'_2$, where we will assume that the indices refer to the same states as before. If we now choose the states "1" and "2" such that in "1" the particle is at rest relative to Σ and in "2" at rest relative to Σ',

* The "inversion" of a relativity-theory equation is the equation one obtains if one replaces all primed quantities with the corresponding unprimed ones and vice versa, replacing also, at the same time, v by $-v$. It follows from the principle of relativity that the equation so obtained must be correct if the original equation was correct.

zu denken wie die entsprechenden ungestrichenen Grössen inbezug
auf Σ. Die Gleichungen (23) lehren nichts neues; sie enthalten das
Additionstheorem der Geschwindigkeiten und sind identisch mit der
Umkehrung der Gleichungen (18). ~~Die~~

Wir würden ein Übergehen der Gleichungen (I) in entsprechend lautende
~~mit ge~~ auch dann erhalten, wenn wir auf den rechten Seiten von (21) und (22)
überall denselben in
~~einen~~ beliebiger Weise von v abhängigen Faktor hinzufügten. Es lässt
sich aber zeigen, dass dieser Faktor gleich 1 sein muss, und zwar in
ganz ähnlicher Weise, wie dies am Schlusse des § 10 für die Lorentz-
Transformation gezeigt wurde.

Die Gleichungen (21) lassen erkennen, dass ~~das~~ dem elektrischen
sowie
~~und~~ dem magnetischen Felde keine besondere Existenz zukommt,
wenn man das Feld von allen berechtigten Systemen aus auffasst. Ein
von Σ aus betrachtet
rein magnetisches Feld besitzt zum Beispiel – von Σ' aus betrachtet – auch
elektrische
~~magnetische~~ Komponenten, wie die zweite und dritte der Gleichungen
(21) zeigt. Das elektromagnetische Feld erscheint somit als ein einheitliches
physikalisches Gebilde, das für jeden Raum-Zeitpunkt durch sechs
Komponenten bestimmt ist. Die formalen Eigenschaften dieses Gebildes werden
später ~~nach Minkowskis Methode bet~~ dargelegt werden.

Gleichung (22) zeigt, dass die elektrische Dichte keine Invariante
Dagegen ist
der Lorentztransformation ist. ~~Wohl aber lässt sich dies für~~ die elektrische
Menge, d. h. das Produkt $\varrho\, d\tau$ invariant, d. h. von der Wahl des Bezugssystems unabhängig. Wählt man nämlich Σ' so, dass es gegenüber dem
betrachteten Volumelement momentan in Ruhe ist, so ist $\mathfrak{q}_x' = 0$, und die
Umkehrung[x] von (22) lautet daher

$$\varrho = \frac{\varrho'}{\sqrt{1-\frac{v^2}{c^2}}}$$

Nach Gleichung (17), ~~ist selbe eine unmittelbare Folge aus der Lorentz-Kontraktion~~
~~ist~~, ist aber

$$d\tau_0 = d\tau' = \frac{d\tau}{\sqrt{1-\frac{v^2}{c^2}}}$$

Aus beiden Gleichungen folgt $\varrho' d\tau' = \varrho\, d\tau$ woraus die Behauptung hervorgeht.
Man kann diese Invarianz auch unmittelbar aus den Gleichungen (I) und
dem Relativitätsprinzip folgern. Aus den ersten beiden Gleichungen (I) folgt
nämlich

$$\operatorname{div}(\mathfrak{q}\varrho) + \frac{\partial\varrho}{\partial t} = 0$$

Diese ~~Satz der~~ Gleichung spricht die Unzerstörbarkeit der elektrischen Menge
aus, denn sie besagt, dass die Zunahme der elektrischen Menge eines Volumenelementes
pro Zeiteinheit gleich ist ~~dem Überschuss der einströmenden über die~~ der in dieses
Zeit in den betrachteten Raum einströmenden Elektrizitätsmenge (Kontinuitätsgleichung). Verfolgen wir also ~~einen Körper~~ ein elektrisches Teilchen (Elektrizitäts-
menge ε) zwischen den Zeiten t_1 und t_2 auf seinem Wege, so ist $\varepsilon_1 = \varepsilon_2$. Dies gilt
berechtigtes
zunächst für ein Bezugssystem K. Für ein zweites berechtigtes Bezugssystem K' gilt analog
$\varepsilon_1' = \varepsilon_2'$, ~~falls~~ wobei wir annehmen wollen, dass sich die Indizes auf die
nämlichen Zustände beziehen wie vorhin. Nehmen wir nun ~~die Anfangs~~
die Zustände „1" und „2" derart, dass bei „1" das Teilchen relativ zu Σ, bei
„2" relativ zu Σ' ruht, so muss nach dem Relativitätsprinzip $\varepsilon_1 = \varepsilon_2'$ sein.
Hieraus folgt die Richtigkeit der Behauptung. Für spätere Betrachtungen

[x] Die „Umkehrung" einer relativitäts-theoretischen Gleichung ist diejenige Gleichung,
die man erhält, wenn man alle gestrichenen Grössen durch die entsprechenden ungestrichenen und umgekehrt, und gleichzeitig v durch $-v$ ersetzt. Dass die so
erhaltene Gleichung richtig sein muss, falls es die ursprüngliche war, folgt aus dem Relativitätsprinzip.
~~und daraus~~

then, according to the relativity principle, we must have $\varepsilon_1 = \varepsilon'_2$. This demonstrates the correctness of the assertion. For later considerations let us note that not only $\rho d\tau$, but consequently, because of (17),

$$\rho_0 = \rho\sqrt{1 - \frac{q^2}{c^2}} \qquad \ldots (24)$$

is also an invariant of the Lorentz transformation, the "rest-density." This is the electrical density for an observer moving with the electricity.

Transformation of the amplitude of an electromagnetic wave. For what follows, it is important to know the law by which the amplitude of a plane (vacuum) wave transforms. Let the wave be given with respect to Σ by the equations

$$\mathfrak{e} = \mathfrak{e}_0 \sin\Phi \qquad \Phi = \omega\left(t - \frac{lx + my + nz}{c}\right),$$
$$\mathfrak{h} = \mathfrak{h}_0 \sin\Phi$$

where $\mathfrak{e}$ and $\mathfrak{h}$ denote spatially constant vectors. Applying the transformation equations (IIb) and (21), we obtain the following equations with respect to Σ'

$$\mathfrak{e}'_x = \mathfrak{e}_{x0} \sin\Phi' \qquad \mathfrak{h}'_x = \mathfrak{h}_{x0} \sin\Phi'$$
$$\mathfrak{e}'_y = b\,(\mathfrak{e}_{y0} - \frac{v}{c}\mathfrak{h}_{z0})\sin\Phi' \qquad \mathfrak{h}'_y = b\,(\mathfrak{h}_{y0} + \frac{v}{c}\mathfrak{e}_{z0})\sin\Phi'$$
$$\mathfrak{e}'_z = b\,(\mathfrak{e}_{z0} + \frac{v}{c}\mathfrak{h}_{y0})\sin\Phi' \qquad \mathfrak{h}'_z = b\,(\mathfrak{h}_{z0} + \frac{v}{c}\mathfrak{e}_{y0})\sin\Phi'$$
$$\Phi' = \omega'\left(t' - \frac{l'x' + m'y' + n'z'}{c}\right)$$

The identity $\Phi = \Phi'$ holds here; the consequences that follow from this identity have already been considered in §11.

We choose the *X*-*Y* plane of Σ to be parallel to the wave normal and consider first the case where the electrical oscillation is parallel to the *Z*-axis. Then, if φ denotes the $\measuredangle$ between the wave normal and the *X*-axis:

$$\mathfrak{e}_{x0} = 0 \qquad \mathfrak{h}_{x0} = -A\sin\varphi$$
$$\mathfrak{e}_{y0} = 0 \qquad \mathfrak{h}_{y0} = -A\cos\varphi$$
$$\mathfrak{e}_{z0} = A \qquad \mathfrak{h}_{z0} = 0$$

From this we obtain for $\mathfrak{e}$ etc. the expressions

$$\mathfrak{e}'_x = 0 \qquad \mathfrak{h}'_x = -A\sin\varphi\sin\Phi'$$
$$\mathfrak{e}'_y = 0 \qquad \mathfrak{h}'_y = b\,(-\cos\varphi + \frac{v}{c})A\sin\Phi'$$
$$\mathfrak{e}'_z = b\,(1 - \frac{v}{c}\cos\varphi)A\sin\Phi' \qquad \mathfrak{h}'_z = 0$$

Thus, the electrical oscillation is also parallel to the z'-axis with respect to Σ' as well. For the amplitude A' of the wave with respect to Σ' we get

$$A' = A\frac{1 - \frac{v}{c}\cos\varphi}{\sqrt{1 - \frac{v^2}{c^2}}}. \qquad \ldots (25)$$

merken wir an, dass nicht nur $\varrho\, d\tau$, sondern infolgedessen wegen (17) auch

$$\varrho_0 = \varrho \sqrt{1 - \frac{q^2}{c^2}} \quad \ldots\ldots (24)$$

eine Invariante der Lorentz-Transformation ist, die „Ruhe-Dichte". Es ist dies nämlich die elektrische Dichte für einen mit der Elektrizität bewegten Beobachter.

Transformation der Amplitude einer elektromagnetischen Welle. Für das Folgende ist es von Bedeutung, das Gesetz zu kennen, nach welchem sich die Amplitude einer ebenen (Vakuum-)Welle transformiert. Die Welle sei inbezug auf Σ durch die Gleichungen gegeben

~~$\mathfrak{e}_x =$ $\mathfrak{e}_y =$ $\mathfrak{e}_z =$~~

$$\mathfrak{e} = \mathfrak{e}_0 \sin \Phi \qquad \Phi = \omega\left(t - \frac{lx + my + nz}{c}\right),$$
$$\mathfrak{h} = \mathfrak{h}_0 \sin \Phi$$

wobei $\mathfrak{e}_0$ und $\mathfrak{h}_0$ räumlich konstante Vektoren bedeuten. Durch Anwendung der Transformationsgleichungen (II b) und (21), erhalten wir inbezug auf Σ' die Gleichungen

$$\mathfrak{e}'_x = \mathfrak{e}_{x0} \sin \Phi' \qquad \mathfrak{h}'_x = \mathfrak{h}_{x0} \sin \Phi'$$
$$\mathfrak{e}'_y = b\left(\mathfrak{e}_{y0} - \frac{v}{c}\mathfrak{h}_{z0}\right) \sin \Phi' \qquad \mathfrak{h}'_y = b\left(\mathfrak{h}_{y0} + \frac{v}{c}\mathfrak{e}_{z0}\right) \sin \Phi'$$
$$\mathfrak{e}'_z = b\left(\mathfrak{e}_{z0} + \frac{v}{c}\mathfrak{h}_{y0}\right) \sin \Phi' \qquad \mathfrak{h}'_z = b\left(\mathfrak{h}_{z0} - \frac{v}{c}\mathfrak{e}_{y0}\right) \sin \Phi'$$
$$\Phi' = \omega'\left(t' - \frac{l'x' + m'y' + n'z'}{c}\right)$$

Dabei gilt die Identität $\Phi = \Phi'$; die aus dieser Identität fliessenden Folgerungen wurden bereits in § 11 behandelt.

Wir wählen die X-Y-Ebene von Σ parallel zur Wellennormale und behandeln zunächst den Fall, dass die elektrische Schwingung parallel zur Z Achse erfolgt. Dann ist, wenn φ den ∢ zwischen Wellennormale und X Achse bedeutet:

$$\mathfrak{e}_{x0} = 0 \qquad \mathfrak{h}_{x0} = -A \sin\varphi$$
$$\mathfrak{e}_{y0} = 0 \qquad \mathfrak{h}_{y0} = -A \cos\varphi$$
$$\mathfrak{e}_{z0} = A \qquad \mathfrak{h}_{z0} = 0$$

Hieraus folgen für $\mathfrak{e}'_x$ etc. die Ausdrücke

$$\mathfrak{e}'_x = 0 \qquad \mathfrak{h}'_x = -A \sin\varphi \sin\Phi'$$
$$\mathfrak{e}'_y = 0 \qquad \mathfrak{h}'_y = b\left(-\cos\varphi + \frac{v}{c}\right) A \sin\Phi'$$
$$\mathfrak{e}'_z = b\left(1 - \frac{v}{c}\cos\varphi\right) A \sin\Phi' \qquad \mathfrak{h}'_z = 0$$

Auch inbezug auf Σ' erfolgt also die elektrische Schwingung parallel zur z'-Achse. Für die Amplitude A' der Welle bezüglich Σ' ergibt sich

$$A' = A \frac{1 - \frac{v}{c}\cos\varphi}{\sqrt{1 - \frac{v^2}{c^2}}} \quad \ldots\ldots (25)$$

Für den Spezialfall, dass der magnetische ~~Kraft senkrecht~~ Feldvektor

The same relation obviously holds for the special case where the *magnetic* field vector is parallel to the *Z*-axis. Since the case of an arbitrary direction of polarization can be constructed from these two special cases by superposition, (25) is generally valid. One can also see that in the case of an arbitrary direction of polarization, the angle between the plane of polarization, on the one hand, and, on the other hand, the plane that is parallel to the *X*-axis and to the wave normal is not changed by the transformation.

§13. Equations of Motion of the Material Point

We have already satisfied ourselves that the equations of motion of classical mechanics are not compatible with the theory of relativity.[56] This presents us with the task of setting up equations of motion for the material point that will satisfy the requirements of the theory of relativity. In order to obtain these equations, we inquire into the law of motion of an electrically charged mass point in an electromagnetic field. For the sake of clarity, we confine ourselves to the case where the accelerating force arises from an electrostatic field that is parallel to the *X*-axis of the coordinate system and where the point moves along the *X*-axis. What is the acceleration of the material point at an arbitrarily chosen space-time point of the motion?

If we denote the reference system to which we refer the process by Σ and the instantaneous velocity of the material point by q, then the point possesses the velocity $q' = 0$ with respect to a reference system Σ' that moves along the *X*-axis of the system Σ with the velocity $v = q$ relative to that system. But Newton's laws of motion are undoubtedly valid for infinitely slow motions. For that reason, for the immediately following moment of time the motion is determined by the equation

$$m \frac{dq'}{dt'} = \varepsilon \mathfrak{e}_x.$$

We only have to transform this equation, which holds for $q' = 0$, to the system Σ in order to obtain the equation of motion that we are seeking, where m is to be viewed as a characteristic constant of the mass point that is not transformed. According to the analysis in the preceding §, ε also retains its value under the transformation. We have, further, by virtue of the first of equations (23) and because we have to set $v = q$,

$$dq' = \frac{dq}{1 - \frac{qv}{c^2}} + \frac{q \frac{v}{c^2} dq}{\left(1 - \frac{qv}{c^2}\right)^2} = \frac{dq}{\left(1 - \frac{q^2}{c^2}\right)^2}.$$

Furthermore, according to the inverse of the fourth of equations (IIb), we get, if we introduce $v = q$ and $\dot{x} = 0$ into the differentiated equation,

$$dt = \frac{dt' + \frac{v}{c^2} dx'}{\sqrt{1 - \frac{v^2}{c^2}}} = \frac{dt'}{\sqrt{1 - \frac{q^2}{c^2}}}.$$

Finally, according to the first of equations (21),

$$\mathfrak{e}'_x = \mathfrak{e}_x.$$

der Z-Achse parallel ist, gilt offenbar dieselbe Beziehung. Da man den Fall beliebiger Polarisationsrichtung aus diesen beiden Spezialfällen durch Superposition konstruieren kann, so gilt (25) allgemein. Man sieht ferner nebenbei, dass im Falle beliebiger Polarisationsrichtung der Winkel zwischen Polarisationsebene einerseits und der Ebene parallel X-Achse und ~~Norma~~ Wellennormale andererseits durch die Transformation nicht geändert wird.

§13. Bewegungsgleichungen des materiellen Punktes.

Wir haben uns bereits davon überzeugt, dass die Bewegungsgleichungen der klassischen Mechanik mit der Relativitätstheorie nicht vereinbar sind. Daraus erwächst für uns die Aufgabe, ~~solche~~ Bewegungsgleichungen für den materiellen Punkt aufzustellen, die den Forderungen der Relativitätstheorie entsprechen. Um zu diesen Gleichungen zu gelangen, fragen wir nach dem Gesetz der Bewegung eines elektrisch geladenen ~~P~~ Massenpunktes in einem elektromagnetischen Felde. Um die Betrachtung recht durchsichtig zu gestalten beschränken wir uns auf den Fall, dass die beschleunigende Kraft von einem der X Achse des Koordinatensystems parallelen elektrostatischen Felde herrührt, und dass sich der Punkt längs der X Achse bewegt. Wie erfolgt die Beschleunigung des materiellen Punktes in einem beliebig gewählten Raum-Zeitpunkte der Bewegung?

Nennen wir Σ das Bezugssystem, auf welches wir den Vorgang beziehen, Σ und q die momentane Geschwindigkeit des materiellen Punktes, so besitzt der Punkt inbezug auf ein Bezugssystem Σ', das ~~man~~ mit der Geschwindigkeit ~~q~~ $v = q$ relativ zu System Σ längs dessen X-Achse bewegt ist gerade die Geschwindigkeit $q' = 0$. ~~Inbezug auf Σ' gilt deshalb~~ Für unendlich langsame Bewegungen ~~dürfen wir~~ gelten aber ohne Zweifel die Newton'schen Bewegungsgesetze. Deshalb ist für das unmittelbar folgende Zeitteilchen die Bewegung durch die Gleichung

$$m \frac{dq'}{dt'} = \varepsilon \mathfrak{X}'_x$$

bestimmt. Wir haben diese Gleichung, welche für $q' = 0$ gilt, nur auf das System Σ zu transformieren, um die gesuchte Bewegungsgleichung zu erhalten, wobei m als eine charakteristische Konstante des Massenpunktes zu betrachten ist, die nicht transformiert wird. Auch ε behält bei der Transformation nach den Betrachtungen des vorigen § seinen Wert. Es ist ferner nach der ersten der Gleichungen (23), und weil ~~$q = 0$~~ $v = q$ zu setzen ist

$$dq' = \frac{dq}{1 - \frac{qv}{c^2}} + \frac{q\frac{v}{c^2}\,dq}{\left(1 - \frac{qv}{c^2}\right)^2} = \frac{dq}{\left(1 - \frac{q^2}{c^2}\right)^2}.$$

~~was wegen~~ Nach der Umkehrung der vierten der Gleichungen (IIb) ist ferner, wenn man in die differenzierte Gleichung $v = q$ und $\dot{x} = 0$ einführt

$$dt = \frac{dt' + \frac{v}{c^2}\,dx'}{\sqrt{1 - \frac{v^2}{c^2}}} = \frac{dt'}{\sqrt{1 - \frac{q^2}{c^2}}}$$

Endlich ist nach der ersten der Gleichungen (21)

$$\mathfrak{X}'_x = \mathfrak{X}_x.$$

Substituting the unprimed quantities for the primed ones in the equation of motion, one obtains first

$$\frac{m\dfrac{dq}{dt}}{\left(1-\dfrac{q^2}{c^2}\right)^{\frac{3}{2}}} = \varepsilon \mathfrak{e}_x \qquad \ldots (26)$$

If one considers the fact that

$$\frac{\dfrac{dq}{dt}}{\left(1-\dfrac{q^2}{c^2}\right)^{\frac{3}{2}}} = \frac{d}{dt}\left\{\frac{q}{\sqrt{1-\dfrac{q^2}{c^2}}}\right\},$$

and that, according to a remark in §2, the right-hand side of (26) is to be viewed as the force $\mathfrak{k}_x$ that acts on the material point, (26) assumes the form

$$\frac{d}{dt}\left\{\frac{mq}{\sqrt{1-\dfrac{q^2}{c^2}}}\right\} = \mathfrak{k}_x$$

Thus, if the law of momentum is to be maintained in the theory of relativity, then the expression inside the curly brackets must be viewed as the momentum of the material point. From this we draw the general conclusion that $\dfrac{m\mathfrak{q}}{\sqrt{1-\dfrac{q^2}{c^2}}}$ is equal to the momentum vector of an arbitrarily moving material point in any arbitrary motion. Thus, if the law of momentum is to be maintained in the theory of relativity and if the foundation of Lorentz's electrodynamics is to be retained, then the vector equation of the motion of the material point under the influence of the arbitrary force $\mathfrak{k}$ must read

$$\frac{d}{dt}\left\{\frac{m\mathfrak{q}}{\sqrt{1-\dfrac{q^2}{c^2}}}\right\} = \mathfrak{k} \qquad \ldots (27)$$

If the only force acting on the material point is of electrodynamical character, then one has to set $\mathfrak{k} = \varepsilon\left\{\mathfrak{e} + \left[\dfrac{\mathfrak{q}}{c}, \mathfrak{h}\right]\right\}$

It can easily be shown that (27) also satisfies the energy law, if $\mathfrak{k}\mathfrak{q}$ is retained as the expression for the work done on the material point per unit time. For one obtains

$$\mathfrak{k}\mathfrak{q} = \mathfrak{q}\frac{d}{dt}\left\{\frac{m\mathfrak{q}}{\sqrt{1-\dfrac{q^2}{c^2}}}\right\} = \frac{d}{dt}\left\{\frac{mq^2}{\sqrt{1-\dfrac{q^2}{c^2}}}\right\} - \frac{m\mathfrak{q}\dot{\mathfrak{q}}}{\sqrt{}} = \frac{d}{dt}\left\{\frac{mq^2}{\sqrt{1-\dfrac{q^2}{c^2}}} + mc^2\sqrt{1-\frac{q^2}{c^2}}\right\}$$

or

$$k\mathfrak{q} = \frac{d}{dt}\left\{\frac{mc^2}{\sqrt{1-\dfrac{q^2}{c^2}}}\right\}. \qquad \ldots (27a)$$

The expression within the brackets on the right plays the role of the energy E of the moving mass point. This expression

$$E = \frac{mc^2}{\sqrt{1-\dfrac{q^2}{c^2}}} \qquad \ldots (28)$$

Ersetzt man in der Bewegungsgleichung die gestrichenen Grössen durch die ungestrichenen, so erhält man zunächst

$$\frac{m\frac{dq}{dt}}{\left(1-\frac{q^2}{c^2}\right)^{\frac{3}{2}}} = \varepsilon \mathfrak{e}_x \quad \ldots \quad (26).$$

Berücksichtigt man, dass

$$\frac{\frac{dq}{dt}}{\left(1-\frac{q^2}{c^2}\right)^{\frac{3}{2}}} = \frac{d}{dt}\left\{\frac{q}{\sqrt{1-\frac{q^2}{c^2}}}\right\}$$

ist, und dass die rechte Seite von (26) nach einer Anmerkung des §2 als die auf den materiellen Punkt wirkende Kraft $\mathfrak{K}_x$ aufzufassen ist, so nimmt (26) die Form an

$$\frac{d}{dt}\left\{\frac{mq}{\sqrt{1-\frac{q^2}{c^2}}}\right\} = \mathfrak{K}_x$$

Soll also in der Relativitätstheorie der Impulssatz aufrecht erhalten werden, so müssen wir den in der geschweiften Klammer stehenden Ausdruck als den Impuls des materiellen Punktes auffassen. Hieraus schliessen wir verallgemeinernd, dass $\frac{m\mathfrak{q}}{\sqrt{1-\frac{q^2}{c^2}}}$ dem Impulsvektor eines beliebig bewegten materiellen Punktes gleich ist. Soll also der Impulssatz in der Relativitätstheorie aufrecht erhalten und die Grundlage der Lorentz'schen Elektrodynamik beibehalten werden, so muss die Vektorgleichung der Bewegung des materiellen Punktes unter der Einwirkung der beliebigen Kraft $\mathfrak{K}$ lauten

$$\frac{d}{dt}\left\{\frac{m\mathfrak{q}}{\sqrt{1-\frac{q^2}{c^2}}}\right\} = \mathfrak{K} \quad \ldots \quad (27)$$

Ist die einzige auf den materiellen Punkt wirkende Kraft elektrodynamischer Natur, so ist hiebei $\mathfrak{K} = \varepsilon\left\{\mathfrak{e} + \left[\frac{\mathfrak{q}}{c}, \mathfrak{h}\right]\right\}$ zu setzen.

Es ist leicht zu zeigen, dass (27) auch dem Energiesatze gerecht wird, wenn ~~der Ausdruck~~ $\mathfrak{K}\mathfrak{q}$ als Ausdruck für die pro Zeiteinheit an dem materiellen Punkte geleistete Arbeit beibehalten wird. Man erhält nämlich

$$\mathfrak{K}\mathfrak{q} = \mathfrak{q}\frac{d}{dt}\left\{\frac{m\mathfrak{q}}{\sqrt{1-\frac{q^2}{c^2}}}\right\} = \frac{d}{dt}\left\{\frac{mq^2}{\sqrt{1-\frac{q^2}{c^2}}}\right\} - \frac{m\mathfrak{q}\dot{\mathfrak{q}}}{\sqrt{}} = \frac{d}{dt}\left\{\frac{mq^2}{\sqrt{1-\frac{q^2}{c^2}}} + mc^2\sqrt{1-\frac{q^2}{c^2}}\right\}$$

oder

$$\mathfrak{K}\mathfrak{q} = \frac{d}{dt}\left\{\frac{mc^2}{\sqrt{1-\frac{q^2}{c^2}}}\right\} \quad \ldots \quad (27a).$$

Der Ausdruck unter der Klammer rechts spielt die Rolle der ~~kinetischen~~ Energie des bewegten Massenpunktes ~~, wobei allerdings … war eine Integrationskonstante~~. Dieser Ausdruck ~~nicht für~~

$$E = \frac{mc^2}{\sqrt{1-\frac{q^2}{c^2}}} \quad \ldots \quad (28)$$

grows to infinity when q approaches the value c; thus, it would require an infinite expenditure of energy to impart the velocity c to the body. In order to see that for small velocities this expression turns into that from Newton's mechanics, we expand the denominator in powers of $\frac{q^2}{c^2}$ and obtain

$$E = mc^2 + \frac{m}{2}q^2 + \ldots \qquad \ldots (28')$$

The second term on the right-hand side is the familiar expression for kinetic energy in classical mechanics. But what does the first, q-independent term signify? To be sure, it does not have, strictly speaking, any legitimacy here; for we have arbitrarily omitted an additive constant in (28). But on the other hand, a glance at (28) shows that the term mc^2 is inseparably linked to the second term of the expansion, $\frac{m}{2}q^2$. One is therefore already inclined at this point to grant a real significance to this term mc^2, to view it as the expression for the energy of the point at rest. According to this conception, we would have to view a body with inertial mass m as an energy store of magnitude mc^2 ("rest-energy" of the body). But we can change the rest-energy of a body, e.g., by supplying heat to it. Thus, if mc^2 is always to be equal to the rest-energy of the body, then the inertial mass m of the body must also change during this warming, with the change amounting to $\frac{\Delta E}{c^2}$, if ΔE denotes the energy increase produced by, e.g., warming. We will show in the next § that this latter consequence—if the theory of relativity is correct—must really hold.

§14. The Inertia of Energy [57]

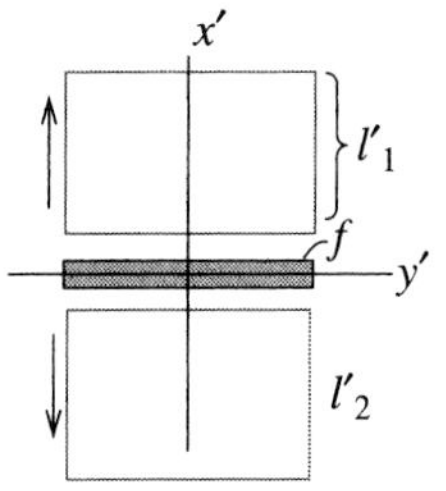

Suppose that a rectangular parallelepiped-shaped plate is at rest relative to Σ', and that its two lateral surfaces, of surface area f, are oriented perpendicularly to the x'-axis. Let these two lateral surfaces send out simultaneously in the positive x' direction and the negative x' direction completely identically constituted wave trains, whose cross-section is f and whose (spatial) lengths, measured in Σ', are equal to $l'_1 = l'_2 = l'$. We imagine that the plate floats freely in space. As a result of the wave emission, the plate will experience completely symmetrical forces owing to the radiation pressure and hence remain at rest.

We wish to determine the energy law and the momentum law for this system, both with respect to Σ' as well as with respect to a system Σ, which is always related to Σ' in the same way as heretofore. To this end, we must first investigate the dimensions and the intensities of the two wave trains with respect to Σ. To begin with, it is obvious that f is also the cross section of the two wave trains with respect to Σ. By contrast, the lengths l_1 and l_2 of the wave trains with respect to Σ are different from l'. For the front plane and rear plane of the one wave train we have, respectively,

$$x' = ct' + \alpha' + l'$$
$$\text{and } x' = ct' + \alpha'.$$

If one transforms by means of equations (IIb), one obtains

wächst ins Unendliche, wenn sich q dem Werte c nähert; es bedürfte also eines unendlichen Energie-Aufwandes, um dem Körper die Geschwindigkeit c zu erteilen. Um zu sehen, dass dieser Ausdruck für kleine Geschwindigkeit in den von Newtons Mechanik übergeht, entwickeln wir den Nenner [nach Potenzen von $\frac{q^2}{c^2}$] und erhalten

$$\mathfrak{E} = mc^2 + \frac{m}{2} q^2 + \cdots \qquad (28')$$

Das zweite Glied ~~m~~ der rechten Seite ist der geläufige Ausdruck der kinetischen Energie der klassischen Mechanik. Was bedeutet aber das erste, von q unabhängige Glied? Dieses hat zwar streng genommen hier keine Legitimation, denn wir haben in (28) eine additive Konstante willkürlich weggelassen. Aber andererseits lehrt ein Blick auf 28), dass das Glied mc^2 mit dem zweiten der Entwicklung $\frac{m}{2} q^2$ untrennbar verbunden ist. Man neigt deshalb schon an dieser Stelle dazu, diesem Term mc^2 eine reale Bedeutung zu geben, ihn als ~~die~~ [den Ausdruck für] die Energie des ruhenden Punktes anzusehen. Nach dieser Auffassung ~~wäre die Masse eines $\mathfrak{E}$ die totale Energie~~ hätten wir einen Körper von der trägen Masse m als einen Energievorrat von der Grösse mc^2 anzusehen („Ruhe-Energie" des Körpers). ~~Ist diese Auffassung zutreffend, so~~ Die Ruhe-Energie eines Körpers können wir aber ändern, z. B. indem wir ihm Wärme zuführen. Soll also mc^2 stets der Ruhe-Energie des Körpers gleich sein, so muss sich bei dieser Erwärmung [die träge Masse m des Körpers] auch ~~m~~ ändern, und zwar um $\frac{\Delta \mathfrak{E}}{c^2}$, wenn mit $\Delta \mathfrak{E}$ der z. B. durch Erwärmung bewirkte Energiezuwachs bezeichnet wird. Dass diese letztere Konsequenz – falls die Relativitätstheorie richtig ist – wirklich zutreffen muss, wollen wir im folgenden § darthun.

§14: Die Trägheit der Energie.

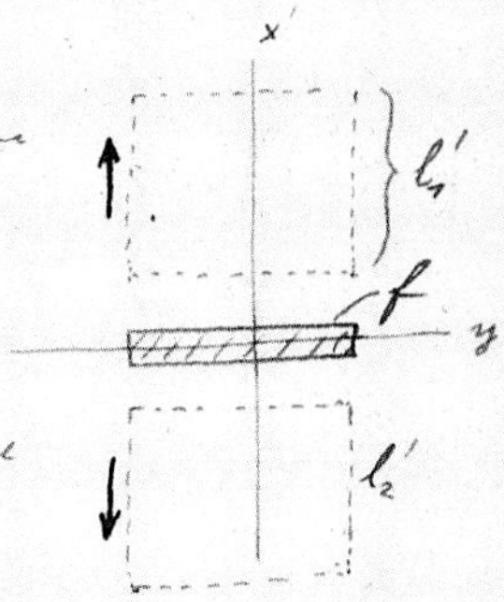

Eine ~~ebene~~ [rechtwinklig-parallelepiped-förmige] Platte ruhe relativ zu Σ' und sei mit ihrem Seitenflächenpaar von der Flächengrösse f senkrecht zur x'-Axe orientiert. Diese beiden Seitenflächen mögen nach der Seite der positiven x' und nach der Seite der negativen x' gleichzeitig vollkommen gleich beschaffene Wellenzüge aussenden, deren Querschnitt f und deren in Σ' gemessene ~~L~~ (räumliche) Länge gleich $l_1' = l_2' = l'$ sei. Die Platte denken wir uns frei im Raume schwebend. Sie wird infolge der Wellenemission vollkommen symmetrische Kräfte durch den Strahlungsdruck erfahren, also in Ruhe bleiben.

Wir wollen für dieses System sowohl inbezug auf Σ' als inbezug auf ein [in derselben Beziehung] wie bisher stets zu Σ' stehendes System Σ den Energiesatz und den Impulssatz aufstellen. Zu diesem Zweck müssen wir zunächst die Abmessungen ~~f~~ [und die Intensitäten] ~~der Sys~~ der beiden Wellenzüge inbezug auf Σ untersuchen. Zunächst ist klar, dass f auch inbezug auf Σ der Querschnitt beider Wellenzüge ist. Hingegen sind die Längen ~~der Wel~~ l_1 und l_2 der Wellenzüge inbezug auf Σ von l' verschieden. Für die Stirnebene bezw. Endebene des einen Wellenzuges ist nämlich

$$x' = ct' + \alpha' + l'$$

bezw. $$x' = ct' + \alpha'$$

Transformiert man mittels der Gleichungen (IIb), so erhält man hieraus:

$$x = ct + \alpha + \frac{l'}{b\left(1 + \frac{v}{c}\right)}$$

and $x = ct + \alpha,$

where, again, we have set $\sqrt{1 - \frac{v^2}{c^2}} = b$, and α denotes another constant. Thus, we obtain for the first wave train with respect to Σ

$$l_1 = l' \sqrt{\frac{1 - \frac{v}{c}}{1 + \frac{v}{c}}}$$

Analogously, we obtain for the second wave train

$$l_2 = l' \sqrt{\frac{1 + \frac{v}{c}}{1 - \frac{v}{c}}}$$

If, furthermore, we denote the amplitudes of the two wave trains with respect to Σ' by A' and the amplitudes with respect to Σ by A_1 and A_2, respectively, then we obtain according to (25)

$$A_1 = A' \sqrt{\frac{1 + \frac{v}{c}}{1 - \frac{v}{c}}} \qquad A_2 = A' \sqrt{\frac{1 - \frac{v}{c}}{1 + \frac{v}{c}}}$$

But the mean density of the electromagnetic energy is equal to $\frac{1}{2}\overline{(\mathfrak{e}^2 + \mathfrak{h}^2)}$ and hence the mean energy density of a plane wave of amplitude A is equal to $\frac{1}{2}\left(\frac{A^2}{2} + \frac{A^2}{2}\right) = \frac{1}{2}A^2$. Hence, if one denotes the energy of each of the two wave trains with respect to Σ' by η', and the energy of the wave trains with respect to Σ by η_1 and η_2, respectively, one obtains

$$\eta' = fl' \cdot \frac{1}{2}A'^2$$

$$\eta_1 = fl_1 \cdot \frac{1}{2}A_1^2 = \frac{1}{2}fl' A'^2 \sqrt{\frac{1 + \frac{v}{c}}{1 - \frac{v}{c}}} \qquad \eta_2 = fl_2 \cdot \frac{1}{2}A_2^2 = \frac{1}{2}fl' A'^2 \sqrt{\frac{1 - \frac{v}{c}}{1 + \frac{v}{c}}},$$

thus

$$\eta_1 + \eta_2 = \frac{2\eta'}{\sqrt{1 - \frac{v^2}{c^2}}} \qquad \text{. . . (29)}$$

In a corresponding way we also calculate the momenta (quantities of motion) I', I_1, and I_2 of the two wave trains. One has to take into account here the fact that the momentum density is $\frac{1}{c}[\mathfrak{e}, \mathfrak{h}]$ (§2), and that the mean momentum density for a plane wave is, thus, equal to $\frac{1}{2c}A^2$. One obtains

$$I' = \frac{1}{2c}fl' A'^2$$

$$I_1 = \frac{1}{2c}fl_1 A_1^2 = \frac{1}{2c}fl' A'^2 \sqrt{\frac{1 + \frac{v}{c}}{1 - \frac{v}{c}}} \qquad I_2 = \frac{1}{2c}fl_2 A_2^2 = \frac{1}{2c}fl' A'^2 \sqrt{\frac{1 - \frac{v}{c}}{1 + \frac{v}{c}}},$$

and thus for the momentum of the two wave trains together with respect to Σ in the direction of the positive x-axis*

$$I_1 - I_2 = \frac{2\eta'}{c^2} \frac{v}{\sqrt{1 - \frac{v^2}{c^2}}} \qquad \text{. . . (30)}$$

* The sum of the momenta of the two wave trains vanishes with respect to Σ'.

$$x = ct + \alpha + \frac{l'}{b\left(1+\frac{v}{c}\right)}$$

$$\text{bezw. } x = ct + \alpha ,$$

wobei wieder $\sqrt{1-\frac{v^2}{c^2}} = b$ gesetzt ist und α eine andere Konstante bedeutet. Es ist also für den ersten Wellenzug inbezug auf Σ

$$~~l_1 = \frac{l'}{b\left(1+\frac{v}{c}\right)}~~ \qquad l_1 = l'\sqrt{\frac{1-\frac{v}{c}}{1+\frac{v}{c}}}$$

Analog ~~erhielte~~ erhält man für den zweiten Wellenzug

$$l_2 = l'\sqrt{\frac{1+\frac{v}{c}}{1-\frac{v}{c}}}.$$

Nennt man ferner A' die Amplitude beider Wellenzüge ~~der Untersuchungen~~ inbezug auf Σ', A_1, bezw. A_2 die Amplituden inbezug auf Σ, so ist nach (25)

$$A_1 = A'\sqrt{\frac{1+\frac{v}{c}}{1-\frac{v}{c}}}$$

$$A_2 = A'\sqrt{\frac{1-\frac{v}{c}}{1+\frac{v}{c}}}$$

Es ist aber $\frac{1}{2}\overline{(\mathfrak{e}^2+\mathfrak{h}^2)}$ gleich der mittleren ~~Energie~~dichte der elektromagnetischen Energie, also die mittlere Energiedichte einer ebenen Welle von der Amplitude A gleich $\frac{1}{2}\left(\frac{A^2}{2}+\frac{A^2}{2}\right) = \frac{1}{2}A^2$. Nennt man also η' die ~~gesamte~~ Energie~~abgabe~~ jedes der beiden Wellenzüge ~~des Systems~~ inbezug auf Σ', ~~so hat man~~ η_1 bezw. η_2 die Energie der Wellenzüge inbezug auf Σ, so hat man

$$\eta' = f l' \cdot \frac{1}{2} A'^2$$

$$\eta_1 = f l_1 \cdot \frac{1}{2} A_1^2 = \frac{1}{2} f l' A'^2 \sqrt{\frac{1+\frac{v}{c}}{1-\frac{v}{c}}}$$

$$\eta_2 = f l_2 \cdot \frac{1}{2} A_2^2 = \frac{1}{2} f l' A'^2 \sqrt{\frac{1-\frac{v}{c}}{1+\frac{v}{c}}},$$

also

$$\eta_1 + \eta_2 = \frac{2\eta'}{\sqrt{1-\frac{v^2}{c^2}}} \qquad (29)$$

In entsprechender Weise berechnen wir auch die Impulse (Bewegungsgrössen) ~~beider~~ J', J_1 bezw. J_2 der beiden Wellenzüge. ~~indem man~~ Dabei ist zu beachten, dass $\frac{1}{c}[\mathfrak{e},\mathfrak{h}]$ die Impulsdichte ist (§2), also für eine ebene Welle die mittlere Impulsdichte gleich $\frac{1}{2c}A^2$. Man erhält

$$J' = \frac{1}{2c} f l' A'^2$$

$$J_1 = \frac{1}{2c} f l_1 A_1^2 = \frac{1}{2c} f l' A'^2 \sqrt{\frac{1+\frac{v}{c}}{1-\frac{v}{c}}}$$

$$J_2 = \frac{1}{2c} f l_2 A_2^2 = \frac{1}{2c} f l' A'^2 \sqrt{\frac{1-\frac{v}{c}}{1+\frac{v}{c}}},$$

also für den Impuls beider Wellenzüge zusammen inbezug auf Σ im Sinne der positiven x Achse[x]

$$J_1 - J_2 = \frac{2\eta'}{c^2}\,\frac{v}{\sqrt{1-\frac{v^2}{c^2}}} \quad \ldots (30)$$

[x] Inbezug auf Σ' verschwindet die Summe der Impulse beider Wellenzüge.

After these preparations we return to our physical problem. As long as the physical state of the plate we are considering (e.g., its energy content) remains unchanged with respect to a comoving observer and we consider exclusively the *translational* motion of the plate as a whole, we surely can view the plate as a material point of a certain mass M. According to (28), one then obtains for the energy E of the plate with respect to Σ, taking into account the circumstance that the plate is at rest with respect to Σ', the expression

$$E = \frac{Mc^2}{\sqrt{1 - \frac{v^2}{c^2}}}.$$

This equation should hold for the state of the plate before the emission of the two light-wave trains. After the emission of the wave trains, the energy of the plate has increased by $-2\eta' = \Delta E'$ with respect to Σ' and by $-(\eta_1 + \eta_2) = \Delta E$ with respect to Σ. On account of (29), the two are related by the equation

$$\Delta E = \frac{\Delta E'}{\sqrt{1 - \frac{v^2}{c^2}}}.$$

According to these two equations, the energy $E + \Delta E$ of the plate after the emission of the wave trains is given by

$$(E + \Delta E) = \frac{\left(M + \frac{\Delta E'}{c^2}\right)}{\sqrt{1 - \frac{v^2}{c^2}}}$$

According to (28), this expression is identical with the expression for the energy of a plate of mass $(M + \frac{\Delta E'}{c^2})$ that is moving with velocity v; *thus, the inertial mass of the plate increases by* $\frac{\Delta E'}{c^2}$ *when its rest-energy (the energy for a comoving observer) experiences an increase of* $\Delta E'$.

It remains for us to show that the same result follows from the momentum conservation law. Before the emission of the wave trains, the momentum B of the plate with respect to Σ is given, according to (27), by

$$B = \frac{Mv}{\sqrt{1 - \frac{v^2}{c^2}}}$$

The change ΔB of the momentum with respect to Σ due to the emission of the wave trains is, according to (30) and according to the momentum law with respect to Σ,

$$\Delta B = -(I_1 - I_2) = \frac{\frac{\Delta E' v}{c^2}}{\sqrt{1 - \frac{v^2}{c^2}}}$$

Thus, the momentum after the emission of the wave trains is

$$(B + \Delta B) = \frac{\left(M + \frac{\Delta E'}{c^2}\right)v}{\sqrt{1 - \frac{v^2}{c^2}}},$$

from which we see again that the inertial mass depends on the rest-energy in the previously indicated manner.

This result is obviously independent of the kind of the processes by which one changes the energy content of a system. It must hold for isolated systems in general;

Nach diesen Vorbereitungen kehren wir zu unserem physikalischen Problem zurück. Solange der physikalische Zustand der von uns betrachteten Platte (z. B. deren Energieinhalt) inbezug auf einen mitbewegten Beobachter ungeändert bleibt, und wir ausschliesslich fortschreitende Bewegung der ~~ganzen~~ Platte ~~ins~~ als Ganzes ins Auge fassen, können wir dieselbe sicherlich als materiellen Punkt von einer gewissen Masse M ansehen. Man erhält dann für die Energie E der Platte inbezug auf Σ unter Berücksichtigung des Umstandes, dass dieselbe inbezug auf Σ' sich in Ruhe befindet, gemäss (28) den Ausdruck

$$E = \frac{Mc^2}{\sqrt{1-\frac{v^2}{c^2}}}$$

Diese Gleichung gelte für den Zustand der Platte vor Aussendung der beiden Licht-wellenzüge. Nach Aussendung der Wellenzüge hat die Energie der Platte inbezug auf Σ' um $-2\eta' = \Delta E'$, inbezug auf Σ um $-(\eta_1 + \eta_2) = \Delta E$ zugenommen. Zwischen beiden besteht wegen (29) die Gleichung

$$\Delta E = \frac{\Delta E'}{\sqrt{1-\frac{v^2}{c^2}}}$$

Nach diesen zwei Gleichungen ist die Energie $E+\Delta E$ ~~nach~~ der Platte nach der Aussendung der Wellenzüge gegeben durch

$$(E+\Delta E) = \frac{\left(M+\frac{\Delta E'}{c^2}\right)}{\sqrt{1-\frac{v^2}{c^2}}}$$

Dieser Ausdruck ist nach (28) gleichlautend wie der Ausdruck der Energie einer mit der Geschwindigkeit v bewegten Platte von der Masse $\left(M+\frac{\Delta E'}{c^2}\right)$; die träge Masse der Platte nimmt also um $\frac{\Delta E'}{c^2}$ zu, wenn ~~sich~~ deren Ruh-Energie (Energie für einen mitbewegten Beobachter) ~~einen Zuwachs~~ um $\Delta E'$ erfährt.

Wir müssen noch zeigen, dass aus dem Impulssatz das nämliche Resultat folgt. Vor der Aussendung der Wellenzüge ist die Bewegungsgrösse B der Platte inbezug auf Σ nach (27) gegeben durch

$$B = \frac{Mv}{\sqrt{1-\frac{v^2}{c^2}}}$$

Die Aenderung ΔB der Bewegungsgrösse inbezug auf Σ infolge Aussendung der Wellenzüge ist nach (30) und nach dem Impulssatze inbezug auf Σ

$$\Delta B = -(J_1 - J_2) = \frac{\frac{\Delta E' v}{c^2}}{\sqrt{1-\frac{v^2}{c^2}}}$$

Also ist die Bewegungsgrösse nach der Aussendung der Wellenzüge

$$B + \Delta B = \frac{\left(M+\frac{\Delta E'}{c^2}\right)v}{\sqrt{1-\frac{v^2}{c^2}}},$$

woraus ~~was~~ wieder hervorgeht, dass die träge Masse in der vorher angegebenen Weise von der Ruh-Energie abhängt. ~~Dies Res~~

~~Wenn die Ar~~ Dies Resultat ist offenbar unabhängig davon, durch was für Vorgänge man den Energieinhalt eines Systems ändert. Es muss allgemein für isolierte Systeme gelten, d. h. für ~~solche~~

i.e., for any system that, like the plate we have just considered, can be treated—taken as a whole—as a freely movable material point.

Unfortunately, this most important of all the consequences of the theory of relativity, which unites the principle of the conservation of mass and the energy principle into *one* principle, is extremely difficult to test experimentally because the value $\frac{\Delta E}{c^2}$ is very small compared with M, for all of the energy changes ΔE that we can make on a system without adding or removing "substance."

Radioactivity might furnish the only direct test of the theorem of the inertia of energy that may prove feasible in the foreseeable future. Thus, the loss of mass that should accompany the complete decay of a gram-molecule of radium (225 g) according to the theory of relativity must be approximately 0.025 g.[58] If the *gravitational* masses measured with a balance are exactly proportional to the *inertial* masses, then the ratio of the atomic weights of radium, on the one hand, and of lead and helium, on the other hand, would have to be reduced by this tiny amount. Of course, such precision in the determination of atomic weights cannot be expected in the foreseeable future.[59]

On the other hand, however, a dependence of mass on energy must be considered very probable when viewed from another standpoint. That is to say, it has been proved experimentally for electrons that their kinetic energy increases with velocity at least approximately in accordance with (28).[60] From this it can be concluded directly that the inertia of an electron gas must increase with the mean kinetic energy of the electrons. It speaks in favor of the theory of relativity that it furnishes such a simple law for this dependence.

jedes System, das ~~in ähnlicher Weise als frei beweglicher materieller~~ wie die vorher betrachtete Platte – als Ganzes aufgefasst – als frei beweglicher materieller Punkt behandelt werden darf.

Leider ist die experimentelle Prüfung dieser wichtigsten aller Konsequenzen aus der Relativitätstheorie, welche das Prinzip von der Erhaltung der Masse mit dem Energieprinzip zu einem Prinzip vereinigt, ungemein schwierig weil für alle Energieänderungen ΔE, die wir an einem System ohne Zuführen oder Wegnehmen von „Substanz" vornehmen können, der Wert $\frac{\Delta E}{c^2}$ gegenüber M sehr klein ist.

Die einzige in absehbarer Zeit vielleicht realisierbare direkte Prüfung des Satzes von der Trägheit der Energie könnte die Radioaktivität liefern. So müsste der Massenverlust, der beim vollständigen Zerfall von einem Grammmolekül Radium (225 g) ~~auftre~~ nach der Relativitätstheorie eintreten müsste, etwa 0,025 g betragen. Falls die mit der Waage gemessenen schweren Massen den trägen Massen streng proportional sind, würde also die Bilanz der Atomgewichte von Radium einerseits, Blei und Helium andererseits um diesen geringen Betrag beeinträchtigt werden. Freilich ist es in absehbarer Zeit nicht zu erhoffen, dass eine solche Präzision in der Bestimmung der Atomgewichte erreicht wird.

Andererseits muss aber von anderen Gesichtspunkten aus eine Abhängigkeit der Masse von der Energie als sehr wahrscheinlich betrachtet werden. Für Elektronen ist nämlich experimentell erwiesen, dass deren kinetische Energie mit der Geschwindigkeit wenigstens annähernd gemäss (28) wächst. Daraus kann unmittelbar gefolgert werden, dass die Trägheit eines Elektronengases mit der mittleren kinetischen Energie der Elektronen wachsen muss. Dass die Relativitätstheorie für ~~dass die Träg~~ diese Abhängigkeit ein so einfaches Gesetz liefert, spricht ~~[illegible]~~ zu ihren Gunsten.

SECTION 3

SOME CONCEPTS AND THEOREMS OF THE FOUR-DIMENSIONAL <GEOMETRY> VECTOR AND TENSOR THEORIES THAT ARE NECESSARY FOR THE COMPREHENSION OF MINKOWSKI'S PRESENTATION OF THE THEORY OF RELATIVITY [61]

§15. The Lorentz Transformation as a Rotational Transformation in Four-Dimensional Space

If, in three-dimensional geometry, a new orthogonal coordinate system with the same coordinate origin is introduced alongside the original orthogonal coordinate system (x, y, z) (rotation of the coordinate system), then the laws of this coordinate transformation are possible from the following two stipulations:

(1) The transformation equations are linear and homogeneous with respect to the coordinates

(2) The distance of an arbitrary point from the coordinate origin is the same with respect to both systems.

For, according to (1), the transformation is determined by equations of the form [62]

$$\begin{aligned} x' &= \alpha_{11} x + \alpha_{12} y + \alpha_{13} z \\ y' &= \alpha_{21} x + \alpha_{22} y + \alpha_{23} z \\ z' &= \alpha_{31} x + \alpha_{31} y + \alpha_{33} z, \end{aligned}$$

where the quantities α are independent of x, y, z. According to (2), these equations must make the equation

$$x^2 + y^2 + z^2 = x'^2 + y'^2 + z'^2$$

into an identity. From this the familiar relations

$$\begin{aligned} \alpha_{11}^2 + \alpha_{21}^2 + \alpha_{31}^2 &= 1 \qquad \alpha_{21}\alpha_{31} + \alpha_{22}\alpha_{32} + \alpha_{23}\alpha_{33} = 0 \\ \alpha_{12}^2 + \alpha_{22}^2 + \alpha_{23}^2 &= 1 \qquad \alpha_{31}\alpha_{11} + \alpha_{32}\alpha_{12} + \alpha_{33}\alpha_{31} = 0 \\ \alpha_{13}^2 + \alpha_{23}^2 + \alpha_{33}^2 &= 1 \qquad \alpha_{11}\alpha_{21} + \alpha_{12}\alpha_{22} + \alpha_{13}\alpha_{23} = 0 \end{aligned}$$

follow. With the help of these equations, by virtue of which only 3 of the constants α can be chosen arbitrarily, one proves directly that the inverse substitution is given by the equations

$$\begin{aligned} x &= \alpha_{11}x' + \alpha_{21}y' + \alpha_{31}z' \\ y &= \alpha_{12}x' + \alpha_{22}y' + \alpha_{32}z' \\ z &= \alpha_{13}x' + \alpha_{23}y' + \alpha_{33}z'. \end{aligned}$$

Thus, both the original and the inverse substitution are completely governed by the table of coefficients

	x	y	z
x'	α_{11}	α_{12}	α_{13}
y'	α_{21}	α_{22}	α_{23}
z'	α_{31}	α_{32}	α_{33}

The coefficients are nothing other than the cosines of the ∠ between the original and the new axes. The essential thing is here that the laws of the transformations in question are completely determined by conditions (1) and (2).—

3. Abschnitt.

Einige ~~Sätz~~ Begriffe und Sätze der vierdimensionalen ~~Geometrie~~ Vektoren- und Tensoren-Theorie, die für das Verständnis von Minkowskis Darstellung der Relativitätstheorie nötig sind.

§15. Die Lorentz-Transformation als ~~Transfo~~ Drehungs-Transformation im vierdimensionalen Raume.

Führt man in der dreidimensionalen Geometrie neben dem ursprünglichen rechtwinkligen Koordinatensystem (x, y, z) ein neues rechtwinkliges Koordinatensystem mit dem nämlichen Anfangspunkt ein (Drehung des Koordinatensystems), so sind die Gesetze dieser ~~Transformation~~ Koordinaten-Transformation ~~ist die durch zwischen den Koordinaten beider Systeme geltenden Gleichungen~~ aus ~~den~~ folgenden zwei Angaben heraus möglich:

1) Die Transformationsgleichungen sind linear und homogen bezüglich der Koordinaten

2) Der Abstand eines beliebigen Punktes vom Koordinaten-Ursprung ist bezüglich beider Systeme der nämliche.

Nach 1) ist nämlich die Transformation durch Gleichungen bestimmt von der Form

$$x' = \alpha_{11} x + \alpha_{12} y + \alpha_{13} z$$
$$y' = \alpha_{21} x + \alpha_{22} y + \alpha_{23} z$$
$$z' = \alpha_{31} x + \alpha_{31} y + \alpha_{33} z,$$

wobei die Grössen α von x, y, z unabhängig sind. Nach 2) müssen durch diese Gleichungen die Gleichung

$$x^2 + y^2 + z^2 = x'^2 + y'^2 + z'^2$$

zu einer Identität machen. Hieraus fliessen die bekannten Relationen

$$\alpha_{11}^2 + \alpha_{21}^2 + \alpha_{31}^2 = 1 \qquad \alpha_{21}\alpha_{31} + \alpha_{22}\alpha_{32} + \alpha_{23}\alpha_{33} = 0$$
$$\alpha_{12}^2 + \alpha_{22}^2 + \alpha_{23}^2 = 1 \qquad \alpha_{31}\alpha_{11} + \alpha_{32}\alpha_{12} + \alpha_{33}\alpha_{31} = 0$$
$$\alpha_{13}^2 + \alpha_{23}^2 + \alpha_{33}^2 = 1 \qquad \alpha_{11}\alpha_{21} + \alpha_{12}\alpha_{22} + \alpha_{13}\alpha_{23} = 0$$

Vermittelst dieser Gleichungen, vermöge ~~welcher~~ nur 3 der Konstanten α willkürlich gewählt werden können, beweist man unmittelbar, dass die inverse Substitution durch die Gleichungen

$$x = \alpha_{11} x' + \alpha_{21} y' + \alpha_{31} z'$$
$$y = \alpha_{12} x' + \alpha_{22} y' + \alpha_{32} z'$$
$$z = \alpha_{13} x' + \alpha_{23} y' + \alpha_{33} z'$$

gegeben ist. Die ursprüngliche, sowie die inverse Substitution wird also durch das Koeffizientenschema

	x	y	z
x'	α_{11}	α_{12}	α_{13}
y'	α_{21}	α_{22}	α_{23}
z'	α_{31}	α_{32}	α_{33}

vollkommen beherrscht. Die Koeffizienten sind nichts anderes als die Kosinus der ✱ zwischen den ursprünglichen und den neuen Axen. Wesentlich ist hiebei, dass die Gesetze der in Betracht kommenden Transformationen durch die Bedingungen 1) und 2) vollkommen bestimmt sind. —

If we compare this with the considerations leading to the general Lorentz transformation from §9, then we see that the transformation equations holding between $x, y, z, u = ict$ and $x', y', z', u' = ict'$ of two justified space-time reference systems satisfy the same conditions and are constructed in the same way as in the just considered three-dimensional case. The only difference is that we now have four coordinates instead of three. We can formulate this in the following way: All of the "justified" time-space reference systems to which the four-dimensional manifold of events is referred are orthogonal coordinate systems with four axes that can be transformed into each other by mere rotation. One has to keep in mind that the fourth coordinate u is always purely imaginary.

The appropriateness of this conception becomes immediately apparent when we consider the special Lorentz transformation from this point of view. The simplest rotational transformation is one that involves only two coordinates. Two cases are possible, depending on whether two spatial coordinates, or one spatial coordinate and the temporal coordinate, undergo a transformation. We put the two cases side by side so as to let their formal equivalence come to the fore. The table of the transformation coefficients takes on the special form

Case 1

	x	y	z	u
x'	α_{11}	α_{12}	0	0
y'	α_{21}	α_{22}	0	0
z'	0	0	1	0
u'	0	0	0	1

Case 2

	x	y	z	u
x'	α_{11}	0	0	α_{14}
y'	0	1	0	0
z'	0	0	1	0
u'	α_{41}	0	0	α_{44}

There are three relations between the still disposable constants α, so that only one more remains that can be disposed of in an arbitrary fashion. If, in both cases, we set $\alpha_{11} = \cos\varphi$, where φ remains arbitrary, then we obtain, by taking into account (16) and by reducing the tables to the transformed coordinates:

Case 1

	x	y
x'	$\cos\varphi$	$\sin\varphi$
y'	$-\sin\varphi$	$\cos\varphi$

Case 2

	x	u
x'	$\cos\varphi$	$\sin\varphi$
u'	$-\sin\varphi$	$\cos\varphi$

In Case 1, φ is a real angle, and the transformation corresponds to a rotation of the coordinate system about the Z-axis or—in terms of the four-dimensional conception—about the z-u plane by the $\measuredangle\ \varphi$.

In Case 2, φ is a purely imaginary angle, because $\cos\varphi$ must be real and $\sin\varphi$ purely imaginary in order for x and x' to be real while u and u' are purely imaginary.

Vergleichen wir hiermit die zur allgemeinen Lorentz-Transformation führenden ~~Entwicklungen~~ Betrachtungen von §9, so sehen wir, dass die Transformationsgleichungen, welche zwischen $x, y, z, u = ict$ und $x', y', z', u' = ict'$ zweier berechtigter raumzeitlicher Bezugssysteme gelten, ~~durch ganz die gleichen~~ gleichen Bedingungen genügen und gleich gebaut sind wie in dem soeben betrachteten dreidimensionalen Falle. ~~Bedingungen beherrscht werden.~~ Der einzige Unterschied ist der, dass wir statt dreier Koordinaten nun deren vier haben. Wir können dies so aussprechen: Die Gesamtheit der berechtigten ~~zeiträumlichen~~ Bezugs-Systeme, ~~mittelst deren~~ auf welche die vierdimensionale Mannigfaltigkeit des Geschehens bezogen werden, sind rechtwinklige vierachsige Koordinatensysteme, die durch blosse Drehung ineinander übergeführt werden können. Es ist dabei nur zu beachten, dass die vierte Koordinate u stets rein imaginär ist.

~~Besonders klar tritt~~ Die ~~Berechtigung~~ Zweckmässigkeit dieser Auffassung tritt sogleich zutage, wenn wir die spezielle Lorentz-Transformation unter diesem Gesichtspunkte betrachten. Die einfachste Drehungs-Transformation ist eine solche von welcher nur zwei Koordinaten betroffen werden. Dabei gibt es zwei Fälle, je nachdem zwei räumliche Koordinaten oder eine räumliche und die zeitliche Koordinate eine Transformation erfahren. Wir stellen beide Fälle neben einander, um ihre formale Gleichwertigkeit hervortreten zu lassen. Die Tabelle der Transformationskoeffizienten ~~geht hier im Speziellen über~~ nimmt die spezielle Form an

1. Fall

	x	y	z	u
x'	α_{11}	α_{12}	0	0
y'	α_{21}	α_{22}	0	0
z'	0	0	1	0
u'	0	0	0	1

2. Fall

	x	y	z	u
x'	α_{11}	0	0	α_{14}
y'	0	1	0	0
z'	0	0	1	0
u'	α_{41}	0	0	α_{44}

Zwischen den noch verfügbaren Konstanten α bestehen drei Relationen, sodass nur mehr eine willkürlich verfügbar bleibt. Setzen wir in beiden Fällen $\alpha_{11} = \cos\varphi$, wobei φ willkürlich bleibt, so erhalten wir, unter Berücksichtigung von (16), indem wir die Tabellen auf die transformierten Koordinaten reduzieren:

1. Fall

	x	y
x'	$\cos\varphi$	$\sin\varphi$
y'	$-\sin\varphi$	$\cos\varphi$

2. Fall

	x	u
x'	$\cos\varphi$	$\sin\varphi$
u'	$-\sin\varphi$	$\cos\varphi$

Im ~~ersten~~ 1. Fall ist φ ein reeller Winkel, und die Transformation ~~ist~~ entspricht einer Drehung des Koordinatensystems um die Z-Axe um den $\measuredangle \varphi$ oder – im Sinne der vierdimensionalen Auffassung – um die Z-U-Ebene. Im ~~zweiten~~ 2. Fall ist φ ein rein imaginärer Winkel, weil $\cos\varphi$ reell, $\sin\varphi$ rein imaginär sein muss, damit x und x' reell, u und u' aber rein imaginär werden. Bezeichnet man mit β eine reelle positive Grösse

Therefore, if β denotes a real positive quantity between 0 and 1, one can set

$$\cos\varphi = \frac{1}{\sqrt{1-\beta^2}} \qquad \sin\varphi = \frac{-i\beta}{\sqrt{1-\beta^2}}$$

In this way one immediately obtains equations (IIa), i.e., the special Lorentz transformation. Thus, the latter corresponds to a rotation of the coordinate system about the *y-z* plane by the (imaginary) angle φ.—

The circumstance that the time coordinate is transformed in transformations from one justified system to another one, and that the time coordinate enters the transformation equations in the same way as the spatial coordinates, led Minkowski to the natural requirement that the mathematical description of physical processes be carried out in such a way that the time coordinate is not distinguished.[63] Thus, instead of the question, "How do physical systems change with time," he poses the question, "How is the four-dimensional structure that consists in the totality of the successive states of a system constituted?"[64] In a manner of speaking, he thus turns the theory of changes (dynamics) in the three-dimensional into a kind of statics of the four-dimensional.[65]

World point. World line. Along with Minkowski, we shall designate the four-dimensional point of the four-dimensional space (space-time point) that corresponds to a specific point-event and is characterized with respect to a justified system by specific coordinate values x, y, z, u, as a "world point." The latter represents the proper element of the mathematical description of nature. If we are concerned, for example, with the description of the motion of a material point, then this is determined if x, y, z are determined as functions of u. In doing so, we define a line in the four-dimensional space, which we shall designate as the "world-line" of the material point. The shape of this line remains unchanged if we replace the reference system with another justified system; only the orientation of the line as a whole relative to the coordinate system changes. The circumstance that the u-coordinates of all of the points of the line are imaginary does not change anything essential here insofar as the general mathematical relationships are concerned.

§16. The Simplest Auxiliary Concepts of Minkowski's Four-Dimensional Theory

In the hitherto prevailing mathematical description of physical processes, in which the time coordinate was distinguished from the spatial coordinates, the equation systems of physics had to have the property of transforming to equation systems of exactly the same form if one introduced a new coordinate system that was at rest relative to the original one, but differently oriented (rotation of the coordinate system). When seeking new laws, it is important therefore to have some means for inspecting an equation or a system of equations for the purpose of determining, without calculation, whether they or it have this property or not. As is well known, these means have been found in the theory of vectors (and its extension, the theory of tensors).[66] At the same time, owing to the charcteristic conciseness of its formulations, this theory

~~zwi~~

zwischen 0 und 1, so kann man daher setzen

$$\cos\varphi = \frac{1}{\sqrt{1-\beta^2}} \qquad \sin\varphi = \frac{-i\beta}{\sqrt{1-\beta^2}}.$$

Man erhält so unmittelbar die Gleichungen (IIa), d. h. die spezielle Lorentz-Transformation. Letztere entspricht also einer Drehung des Koordinatensystems ~~um~~ die y-z-Ebene um den (imaginären) Winkel φ. —

Der Umstand, dass die Zeit-Koordinate sich bei Transformationen von einem berechtigten System auf ein anderes transformiert, und dass die Zeitkoordinate in die Transformationsgleichungen in gleicher Weise eingeht wie die räumlichen Koordinaten, führte Minkowski ~~dazu~~ zu der natürlichen Forderung, die mathematische Beschreibung physikalischer Vorgänge sei so durchzuführen, dass dabei die Zeitkoordinate nicht ausgezeichnet wird. An die Stelle der Frage „Wie ändern sich die physikalischen Systeme mit der Zeit?" setzt er also die Frage „Wie ~~ist~~ ~~sieht das ges~~ ist das vierdimensionale ~~Gesamtheit aller Zustände~~ Gebilde beschaffen, welches in der Gesamtheit aller aufeinanderfolgenden Zustände eines Systems besteht?" Er verwandelt damit gewissermassen die Lehre von den Veränderungen (Dynamik) im Dreidimensionalen in eine Art Statik des Vierdimensionalen.

Weltpunkt. Weltlinie. Mit Minkowski wollen wir den vierdimensionalen Punkt des vierdimensionalen ~~Punktes~~ Raumes (Raum-Zeit-Punkt), der einem bestimmten ~~Rau~~ Punktereignis entspricht und inbezug auf ein berechtigtes ~~System~~ Bezugssystem durch bestimmte Koordinatenwerte x, y, z, u gekennzeichnet ist, als „Weltpunkt" bezeichnen. Dieser stellt das eigentliche Element der mathematischen Naturbeschreibung dar. Handelt es sich z. B. um die Beschreibung der Bewegung eines materiellen Punktes, so ist diese bestimmt wenn x, y, z als Funktion von ~~t bzw.~~ u bestimmt sind. Hierdurch wird eine Linie im vierdimensionalen Raume definiert, ~~deren~~ welche wir als „Weltlinie" des materiellen Punktes zu bezeichnen haben. Die Gestalt dieser Linie bleibt ungeändert, wenn wir das Bezugssystem durch ein anderes gleichberechtigtes ersetzen; es ändert sich nur die Orientierung der Linie als Ganzes relativ zum Koordinatensystem. Der Umstand, dass ~~sam~~ die u-Koordinaten sämtlicher Punkte der Linie imaginär sind, ändert hieran nichts Wesentliches, insoweit es sich um die allgemeinen mathematischen Zusammenhänge handelt.

§ 16. ~~Der Vierervektor.~~ Die einfachsten Hilfsbegriffe von Minkowskis vierdimensionaler Theorie.

~~Bei~~ der bisherigen ~~physikalischen~~ mathematischen Beschreibung physikalischer Vorgänge, ~~hat es sich als zweckmässig erwiesen,~~ ~~derartige~~ ~~solche~~ ~~Verknüpfungen von Grössen die in den Gle~~ bei welcher die Zeitkoordinate gegenüber den räumlichen Koordinaten ausgezeichnet wurde, mussten die Gleichungssysteme der Physik die Eigenschaft besitzen, bei ~~Aenderung~~ Einführung eines neuen Koordinatensystems, das gegenüber dem ursprünglichen ruhte, aber anders orientiert war (Drehung des Koordinatensystems), in Gleichungssysteme von genau derselben Form überzugehen. Beim Aufsuchen neuer Gesetze ist es daher von Wichtigkeit, Mittel zu besitzen, um es ~~einer~~ Gleichung bzw. einem Gleichungssystem ohne Rechnung ansehen zu können, ob es diese Eigenschaft ~~zeigt~~ hat oder nicht. Diese Mittel wurden bekanntlich in der Vektorentheorie (und ihrer Erweiterung, der Theorie der Tensoren) gefunden. Diese Theorie erhöht gleichzeitig durch die ihr eigene Kürze der Formulierung die Übersicht über die Gleichungssysteme.

provides for an enhanced overview of the systems of equations. The vector calculus attains its goal by uniting under one concept three quantities (vector components), whose transformation properties are identical with those of the coordinates of a point (and thus known once and for all), and denoting them by a common symbol.

Minkowski had the very fruitful idea of reshaping the equations of the theory of relativity, which include, with his choice of the time coordinate, four equivalent coordinates that are wholly analogous to the coordinates x, y, z of spatial geometry, in a manner very similar to the way this is done by the vector calculus with respect to three-dimensional space. He accomplishes that (following the example of vector analysis) by conceptionally uniting a number of quantities (or differential operations) whose transformation properties are of a certain type. In what follows, we shall consider the most important of these auxiliary concepts that greatly simplify the system of the theory of relativity.

~~*Four-*~~*vector.* [67] The simplest auxiliary concept of this kind corresponds to the (axial) vector of ~~three-dimensional~~ vector analysis. [68] Such a vector ($\mathfrak{a}$) is the combination of three quantities $\mathfrak{a}_x$, $\mathfrak{a}_y$, $\mathfrak{a}_z$ ~~that~~ ~~transform~~ like the coordinates x, y, z of a spatial point. In Minkowski's four-dimensional space (~~called~~ "world"), the four-vector A corresponds to this structure.* This four-vector represents the ~~aggregate~~ of four quantities, A_x, A_y, A_z, A_u, which are referred to as its components; A_x, A_y, A_z are always real and A_u is purely imaginary. The basic property of the four-vector is that in a general Lorentz transformation its components transform like the coordinates x, y, z, u of a world-point, that is, according to the schema

	A_x	A_y	A_z	A_u
A'_x	α_{11}	α_{12}	α_{13}	α_{14}
A'_y	α_{21}	α_{22}	α_{23}	α_{24}
A'_z	α_{31}	α_{32}	α_{33}	α_{34}
A'_u	α_{41}	α_{42}	α_{43}	α_{44}

~~in~~ accordance with the equations

$$A' = \alpha_{11}A_x + \alpha_{12}A_y + \alpha_{13}A_z + \alpha_{14}A_u \text{ etc.}$$

and

$$A_x = \alpha_{11}A'_x + \alpha_{21}A'_y + \alpha_{31}A'_z + \alpha_{41}A'_u \text{ etc.}$$

For the special Lorentz transformation (uniform ~~translation~~ in the direction of the X-axis without *spatial* rotation) we have the equations

$$\left.\begin{aligned} A'_x &= \frac{A_x + i\beta A_u}{\sqrt{1-\beta^2}} \\ A'_y &= A_y \\ A'_z &= A_z \\ A'_u &= \frac{A_u - i\beta A_x}{\sqrt{1-\beta^2}} \end{aligned}\right\} \quad (31)$$

* Along with Laue, we shall denote such four-vectors by capital letters of the Greek alphabet. [69]

Die Vektorrechnung erreicht ihr Ziel dadurch, dass sie drei Grössen (Vektorkomponenten) deren Transformationseigenschaften mit denen der Koordinaten eines ~~[illegible]~~ Punktes übereinstimmen (also ein für allemal bekannt sind) ~~zu einem Begriff~~ zu einem Begriff vereinigt und ~~den Inbegriff~~ mit einem gemeinsamen Zeichen bezeichnet.

Minkowski hatte nun die sehr fruchtbare Idee, ~~zu~~ die Gleichungen der ~~[illegible]~~ Relativitätstheorie, in welche bei seiner Wahl der Zeitkoordinate ~~[illegible]~~ vier gleichwertige Koordinaten eintreten, die den Koordinaten x, y, z der räumlichen Geometrie ganz analog sind, in ganz ähnlicher Weise umzuformen, wie es die Vektorrechnung inbezug auf den dreidimensionalen Raum leistet. Er leistet dies (nach dem Beispiel der Vektoranalysis), indem er eine Anzahl Grössen (bezw. Differenzialoperationen), ~~zu einer begrifflichen Einheit~~ deren Transformationseigenschaften ~~ein für allemal untersucht werden~~ bestimmten Typen angehören, begrifflich zusammenfasst. Die wichtigsten dieser das System der Relativitätstheorie ungemein vereinfachenden Hilfsbegriffe wollen wir im Folgenden ~~[illegible]~~ betrachten.

<u>Vierervektor</u>. Der einfachste derartige Hilfsbegriff entspricht dem (axialen) Vektor der dreidimensionalen Vektoranalysis. Ein solcher Vektor (v) ist die Zusammenfassung dreier Grössen v_x, v_y, v_z, die sich so transformieren wie die Koordinaten x, y, z eines Raumpunktes. Diesem Gebilde entspricht in Minkowskis vierdimensionalem Raum („Welt" genannt) der Vierervektor* A. Er ~~besitzt vier Komponenten~~ bedeutet den Inbegriff von vier Grössen A_x, A_y, A_z, A_u, die als seine Komponenten bezeichnet werden. Dabei sind A_x, A_y, A_z stets reell, A_u ist rein imaginär. Die Grundeigenschaft des Vierervektors ist die, dass sich seine Komponenten bei einer allgemeinen Lorentz-Transformation so transformieren wie die Koordinaten x, y, z, u eines Weltpunktes, also nach dem Schema

	A_x	A_y	A_z	A_u
A'_x	α_{11}	α_{12}	α_{13}	α_{14}
A'_y	α_{21}	α_{22}	α_{23}	α_{24}
A'_z	α_{31}	α_{32}	α_{33}	α_{34}
A'_u	α_{41}	α_{42}	α_{43}	α_{44}

gemäss den Gleichungen

$$A'_x = \alpha_{11} A_x + \alpha_{12} A_y + \alpha_{13} A_z + \alpha_{14} A_u \quad \text{etc.}$$

und

$$A_x = \alpha_{11} A'_x + \alpha_{21} A'_y + \alpha_{31} A'_z + \alpha_{41} A'_u \quad \text{etc.}$$

~~Bei~~ Für die speziellen Lorentz-Transformation (Gleichförmige Translation in Richtung der X-Achse ohne <u>räumliche</u> Drehung) gelten die Gleichungen

$$\left.\begin{aligned} A'_x &= \frac{A_x + i\beta A_u}{\sqrt{1-\beta^2}} \\ A'_y &= A_y \\ A'_z &= A_z \\ A'_u &= \frac{A_u - i\beta A_x}{\sqrt{1-\beta^2}} \end{aligned}\right\} \qquad (31)$$

---

* Wir wollen solche Vierervektoren ~~mit den~~ von Lauenmit den grossen Buchstaben des griechischen Alphabetes bezeichnen.

Four-vector.[70] Let a structure be determined with respect to the (justified) reference system Σ by the four quantities A_1, A_2, A_3, A_4 associated with the four coordinate axes. With respect to another, arbitrarily chosen reference system Σ', let the structure be determined by four other quantities A'_1, A'_2, A'_3, A'_4. Following Minkowski, one calls the structure a four-vector if the transformation relations between A'_1, A'_2, etc., and the corresponding unprimed quantities are the same as those between the quantities

$$\begin{aligned} x'_1 &= x' \\ x'_2 &= y' \\ x'_3 &= z' \\ x'_4 &= ict' = u' \end{aligned}$$

and the corresponding quantities of the unprimed system. We call the quantities x_ν (four-dimensional) point coordinates, and the quantities A_ν the components of the four-vector. Thus, according to §9, we have the equations

$$x'_\mu = \sum \alpha_{\mu\nu} x_\nu \qquad \text{. . . (31)}$$

$$A'_\mu = \sum_\nu \alpha_{\mu\nu} A_\nu \qquad \text{. . . (32)}$$

We represent the four-vector as a whole by the symbol (A_ν).[71] The three-dimensional vector quantities along with their components will always be denoted by lower-case letters, and the four-dimensional ones by upper-case letters. Thus, equation (32) is the definition of the four-vector.

Tensor (second rank). Let there be given two four-vectors (A_ν) and (B_ν). We denote the individual products of the form $A_\mu B_\nu$ by $T_{\mu\nu}$. According to (32), the following transformation equations will hold, then, for these products $T_{\mu\nu}$

$$T'_{\mu\nu} = \sum_{\sigma\tau} \alpha_{\mu\sigma} \alpha_{\nu\tau} T_{\sigma\tau} \qquad \text{. . . (33)}$$

In general, we designate a structure that <with respect to a coordinate system reference system Σ> is defined by ten quantities $T_{\sigma\tau}$, each of which is associated with two coordinate axes of Σ, as a *tensor* (second-rank) if the transformation equations (33) hold for the quantities $T_{\sigma\tau}$. We call the quantities $T_{\sigma\tau}$ the components of the tensor. It should not be assumed that the components $T_{\sigma\tau}$ and $T_{\tau\sigma}$ must be identical to one another, even though that was the case in the example that led us to (33).[72] We represent the structure as a whole, and thus the tensor itself, by the symbol $(T_{\sigma\tau})$.

Tensor of the n-th rank.[73] The structure defined by 4^n quantities $T_{\sigma_1\sigma_2\ldots\sigma_n}$ that are associated with the coordinate axes we call an *n*th-rank tensor $(T_{\sigma_1\ldots\sigma_n})$ if these quantities obey the transformation equations

$$T'_{\tau_1\ldots\tau_n} = \sum_{\sigma_1\ldots\sigma_n} \alpha_{\tau_1\sigma_1}\alpha_{\tau_2\sigma_2}\ldots\alpha_{\tau_n\sigma_n} T_{\sigma_1\ldots\sigma_n} \qquad \text{. . . (34)}$$

We call the quantities $T_{\sigma_1\ldots\sigma_n}$ the components of the tensor.—

One can see from (32) and (34) that the four-vector is nothing other than a first-rank tensor.

Vierervektor. Ein Gebilde sei durch vier den Koordinatenachsen zugeordnete Grössen A_1, A_2, A_3, A_4 inbezug auf das (berechtigte) Bezugssystem Σ bestimmt. ~~bezüglich inbezug auf irgend~~ Inbezug auf ein beliebig~~es~~ gewähltes anderes berechtigtes Bezugssystem Σ' sei es durch durch vier andere Grössen A_1', A_2', A_3', A_4' bestimmt. Man nennt das Gebilde nach Minkowski einen Vierervektor, wenn zwischen A_1', A_2' etc und ~~den~~ den entsprechenden ungestrichenen Grössen dieselben Transformations-Beziehungen existieren wie zwischen den Grössen

$$x_1' = x'$$

$$x_2' = y'$$

$$x_3' = z'$$

$$x_4' = ict' = u'$$

und den entsprechenden Grössen des ungestrichenen Systems. Die Grössen x_ν nennen wir (vierdimensionale) Punktkoordinaten, die Grössen A_ν die Komponenten des Vierervektors. ~~Mit Hilfe der Betrachtungen des~~ Nach § 9 bestehen also die Gleichungen

$$x_\mu' = \sum \alpha_{\mu\nu} x_\nu \quad \ldots\ldots \quad (31)$$

$$A_\mu' = \sum_\nu \alpha_{\mu\nu} A_\nu \quad \ldots\ldots \quad (32)$$

Den Vierervektor als Ganzes ~~bezeichnen~~ stellen wir durch das Zeichen (A_ν) dar. Vierdimensionale Vektorgrössen nebst deren Komponenten wollen wir stets mit kleinen, vierdimensionale mit grossen Buchstaben bezeichnen. Gleichung (32) ist nach dem Gesagten die Definition des Vierervektors.

Tensor (zweiten Ranges) Es seien zwei Vierervektoren (A_μ) und (B_ν) gegeben. ~~Wir können die Kombination dieser beiden wieder als ein Gebilde auffassen und das Gebilde~~ Wir bezeichnen mit $T_{\mu\nu}$ die einzelnen Produkte von der Form $A_\mu B_\nu$. Nach (32) bestehen dann für diese Produkte $T_{\mu\nu}$ die Transformationsgleichungen

$$T_{\mu\nu}' = \sum_{\sigma\tau} \alpha_{\mu\sigma} \alpha_{\nu\tau} T_{\sigma\tau} \quad \ldots\ldots \quad (33)$$

Wir bezeichnen nun allgemein ein Gebilde, welches ~~bezüglich eines Koordinatensystems Bezugssystems~~ Σ durch 16 je zwei Koordinatenachsen von Σ zugeordnete Grössen $T_{\sigma\tau}$ definiert ist, dann als einen Tensor (zweiten Ranges), wenn für die Grössen $T_{\sigma\tau}$ die Transformationsgleichungen (33) bestehen. ~~Wir brauchen~~ Die Grössen $T_{\sigma\tau}$ nennen wir die Komponenten des Tensors. ~~dabei~~ Es sei dabei nicht vorausgesetzt, dass die Komponenten $T_{\sigma\tau}$ und $T_{\tau\sigma}$ einander gleich sein müssen, obwohl dies bei dem Beispiel zutraf, welches uns auf (33) geführt hat. Das Gebilde als Ganzes, also den Tensor selbst, stellen wir durch das Zeichen $(T_{\sigma\tau})$ dar.

Tensor n-ten Ranges. Das durch 4^n ~~Grössen~~ den Koordinatenachsen zugeordnete Grössen $T_{\sigma_1\sigma_2\ldots\sigma_n}$ definierte Gebilde nennen wir dann einen Tensor n-ten Ranges $(T_{\sigma_1\ldots\sigma_n})$, wenn für diese Grössen die Transformationsgleichungen

$$T_{\tau_1\ldots\tau_n}' = \sum_{\sigma_1\ldots\sigma_n} \alpha_{\tau_1\sigma_1} \alpha_{\tau_2\sigma_2} \cdots \alpha_{\tau_n\sigma_n} T_{\sigma_1\ldots\sigma_n} \quad \ldots\ldots \quad (34)$$

gelten. Die Grössen $T_{\sigma_1\ldots\sigma_n}$ nennen wir die Komponenten des Tensors. —
Man sieht aus (32) und (34), dass der Vierervektor nichts anderes ist als ein Tensor ersten Ranges.

All covariants introduced into Minkowski's theory can easily be based on tensors. Two kinds of special tensors are of great importance, which we will now define.

Symmetric Tensor.

If $(T_{\sigma\tau})$ is a second-rank tensor, then, as a glance at (33) shows, $(T_{\tau\sigma})$ is also a second-rank tensor. But these two tensors do not coincide with each other, because in general $T_{\sigma\tau}$ is different from $T_{\tau\sigma}$. For that reason we obtain a second-rank tensor of a special kind if we stipulate that for every combination of indices we should have

$$T_{\sigma\tau} = T_{\tau\sigma}.$$

We call such a tensor a "symmetric" tensor of the second rank. This tensor possesses not 16, but only 10 components that differ from one another.

As a quite special case of such a tensor, let me mention one, which we shall denote by $(\delta_{\sigma\tau})$, with the components

$$\begin{array}{cccc} 1 & 0 & 0 & 0 \\ 0 & 1 & 0 & 0 \\ 0 & 0 & 1 & 0 \\ 0 & 0 & 0 & 1 \end{array}$$

The fact that this is a tensor follows directly from (33) in conjunction with the relations obtaining between the α.—

In general, we call the tensor $(T_{\sigma_1 \ldots \sigma_n})$ symmetric if all the components that are obtained from one another by an interchange of the indices are identical to one another. If we wish to indicate that a tensor $(T_{\sigma_1 \ldots \sigma_n})$ is symmetric, we signify this by a horizontal bar, and thus write: $(\overline{T_{\sigma_1 \ldots \sigma_n}})$ [74]

Vector [75]

If the components $T_{\sigma\tau}$ of a second-rank tensor [76] $(T_{\sigma\tau})$ satisfy the condition

$$T_{\sigma\tau} = - T_{\sigma\tau},$$

then we call the tensor a second-rank vector. Due to the disappearance of the components with two identical indices, the latter has 12 components, which, however, because of the relation defining them, have, pair-wise, the same magnitude but opposite signs; for that reason, following Sommerfeld, one usually calls this tensor a "six-vector." [77]

In general, we call a tensor $(T_{\sigma_1 \ldots \sigma_n})$ a vector if two of its components that are obtained from each other by an interchange of the indices always have the same magnitude but opposite signs. <It will turn out that in a four-dimensional continuum only the six-vector and a very special vector of rank four, which we will now consider, are of importance.>

When it seems desirable to do so, we will signify the tensor character of a vector by a vertical bar, i.e., we will write: $(\overset{|}{T}_{\sigma_1 \ldots \sigma_n})$.[78]

Auf den Tensor lassen sich alle in Minkowskis Theorie eingeführten speziellen Kovarianten leicht zurückführen. Es gibt zwei Arten von Tensoren von besonderer Wichtigkeit, die wir nun definieren.

Symmetrischer Tensor.

Wenn ($T_{\sigma\tau}$) ein Tensor zweiten Ranges ist, so ist, wie ein Blick auf (33) lehrt, auch ($T_{\tau\sigma}$) ein Tensor zweiten Ranges. Aber es stimmen diese beiden Tensoren nicht miteinander überein, weil im Allgemeinen $T_{\sigma\tau}$ von $T_{\tau\sigma}$ verschieden ist. Einen ~~speziellen~~ Tensor zweiten Ranges spezieller Art erhalten wir daher, wenn wir festsetzen, dass für jede Index-Kombination

$$T_{\sigma\tau} = T_{\tau\sigma}$$

sein soll. Wir nennen einen solchen Tensor einen „symmetrischen" Tensor zweiten Ranges. Er besitzt nicht 16, sondern nur 10 voneinander verschiedene Komponenten.

Als ganz speziellen Fall eines solchen Tensors erwähne ich einen, den wir mit ($S_{\sigma\tau}$) bezeichnen wollen mit den Komponenten

$$\begin{matrix} 1 & 0 & 0 & 0 \\ 0 & 1 & 0 & 0 \\ 0 & 0 & 1 & 0 \\ 0 & 0 & 0 & 1 \end{matrix}$$

Dass dies ein Tensor ist, folgt unmittelbar aus (33) in Verbindung mit den zwischen den α bestehenden Relationen.—

Allgemein nennen wir den Tensor ($T_{\sigma_1 \ldots \sigma_n}$) einen symmetrischen, wenn alle Komponenten, die durch Vertauschung ~~zweier seiner~~ der Indizes aus einander hervorgehen, einander gleich sind. Wollen wir angeben, dass ein Tensor ($T_{\sigma_1 \ldots \sigma_n}$) ein symmetrischer ist, so ~~wollen~~ deuten wir dies durch einen wagrechten Strich an, schreiben also: ($\overline{T}_{\sigma_1 \ldots \sigma_n}$)

~~Vektor~~.

Vektor

Wenn die Komponenten $T_{\sigma\tau}$ ~~ein~~ Tensor zweiten Ranges ($T_{\sigma\tau}$) die Bedingung

$$T_{\sigma\tau} = -T_{\tau\sigma}$$

erfüllen, so nennen wir den Tensor einen Vektor zweiten Ranges. Derselbe hat wegen des Verschwindens der ~~Glieder~~ Komponenten mit zwei gleichen Indizes 12 Komponenten, die aber wegen der Definitionsbeziehung paarweise entgegengesetzt gleich sind. deswegen wird er nach Sommerfeld gewöhnlich als „Sechservektor" bezeichnet.

Allgemein nennen wir einen Tensor ($T_{\sigma_1 \ldots \sigma_n}$) dann einen Vektor, wenn zwei seiner Komponenten, die durch Vertauschung zweier Indizes aus einander hervorgehen, einander stets entgegengesetzt gleich sind. ~~Wir wollen den Vektorcharakter durch einen über die Tensorbezeichnung~~ ~~Es wird sich zeigen, dass in einem Kontinuum von vier Dimensionen nur der Sechservektor sowie ein ganz spezieller Vektor vierten Ranges von Bedeutung ist, den wir sogleich betrachten wollen.~~

Den Tensorcharakter eines Vektors wollen wir, wenn es wünschbar erscheint, durch einen vertikalen Strich bezeichnen, also schreiben: ($\overset{|}{T}_{\sigma_1 \ldots \sigma_n}$).

Nach dieser Definition haben wir eigentlich kein Recht, den Tensor

According to this definition, we have in fact no right to designate the tensor of rank one as a "vector" (four-vector), because an interchange of indices is out of the question here. But it will do no harm if we use this familiar expression all the same.

Since components of vectors with two identical indices vanish, it is clear that in the case of our four-dimensional manifold, vectors of rank three have only four different components, neglecting the differences in sign, vectors of rank four only *one* component, neglecting the sign, and that vectors of a rank higher than four do not exist.

Consequently, vectors of rank four can be replaced by a specfic $(e_{\sigma_1 \ldots \sigma_4})$, which is defined as follows: We have

$$e_{\sigma_1 \ldots \sigma_4} = \pm 1,$$

depending on whether the permutation σ_1, σ_2, σ_3, σ_4 is obtained from the permutation 1, 2, 3, 4 by an even or by an odd number of interchanges of two indices. If $(e_{\sigma_1 \ldots \sigma_4})$ is a *tensor*, then according to this definition it is a *vector*; that it is a tensor can easily be proved by using the familiar laws of determinants.[79]

§17. Addition and Multiplication of Tensors

Addition and subtraction of tensors

It follows from (34) that one again obtains the components of a tensor if one adds or subtracts the corresponding components of two tensors of the same rank. One speaks in this sense of the addition or subtraction of tensors. Thus, one can write

$$(T_{\sigma_1 \ldots \sigma_n}) \pm (U_{\sigma_1 \ldots \sigma_n}) = (T_{\sigma_1 \ldots \sigma_n} \pm U_{\sigma_1 \ldots \sigma_n}) \qquad \ldots (35)$$

For the sum or difference, symmetric tensors here again yield symmetric tensors, and vectors again yield vectors.

Remark. We are able to form a symmetric tensor and a vector of rank two for every tensor of rank two. If $(T_{\sigma\tau})$ is the given tensor, then, according to (33), $(T_{\tau\sigma})$ is also a tensor, and hence

$(T_{\sigma\tau} + T_{\tau\sigma})$ is a symmetric tensor of rank two

$(T_{\sigma\tau} - T_{\tau\sigma})$ is a vector of rank two

Outer Products of Tensors

From two tensors with the components $T_{\sigma_1 \ldots \sigma_m}$ and $U_{\tau_1 \ldots \tau_n}$ one can form a new tensor with the components

$$T_{\sigma_1 \ldots \sigma_m} U_{\tau_1 \ldots \tau_n}.$$

We call the tensor $(T_{\sigma_1 \ldots \sigma_m} U_{\tau_1 \ldots \tau_n})$ the outer product of the two given tensors. For according to (34)

$$T'_{s_1 \ldots s_m} = \sum_{\sigma_1 \ldots \sigma_m} \alpha_{s_1\sigma_1} \alpha_{s_2\sigma_2} \ldots \alpha_{s_m\sigma_m} T_{\sigma_1 \ldots \sigma_m}$$

$$U'_{t_1 \ldots t_n} = \sum_{\tau_1 \ldots \tau_n} \alpha_{t_1\tau_1} \alpha_{t_2\tau_2} \ldots \alpha_{t_n\tau_n} U_{\tau_1 \ldots \tau_n},$$

from which one obtains by multiplication

$$T'_{s_1 \ldots s_m} U'_{t_1 \ldots t_n} = \sum_{\sigma_1 \ldots \sigma_m \tau_1 \ldots \tau_n} \alpha_{s_1\sigma_1} \ldots \alpha_{s_m\sigma_m} \alpha_{t_1\tau_1} \ldots \alpha_{t_n\tau_n} T_{\sigma_1 \ldots \sigma_m} U_{\tau_1 \ldots \tau_n}.$$

ersten Ranges als einen „Vektor" (Vierervektor) zu bezeichnen, indem hier von einer Vertauschung der Indizes nicht die Rede sein kann. Es wird aber kein Schaden daraus entstehen, wenn wir uns doch dieses geläufigen Ausdruckes bedienen.

Da Komponenten von Vektoren mit zwei gleichen Indizes verschwinden, so ist klar, dass im Falle unserer vierdimensionalen Mannigfaltigkeit Vektoren dritten Ranges abgesehen von Vorzeichenunterschieden nur vier verschiedene Komponenten haben, dass Vektoren vierten Ranges abgesehen vom Vorzeichen nur eine Komponente haben, und dass Vektoren von höherem als dem vierten Range nicht existieren.

Die Vektoren vierten Ranges lassen sich infolgedessen durch einen bestimmten ~~Vektor~~ $(e_{\sigma_1 \ldots \sigma_4})$ ersetzen, der folgendermassen definiert ist: Es ist

$$e_{\sigma_1 \ldots \sigma_4} = \pm 1,$$

je nachdem die Permutation $\sigma_1, \sigma_2, \sigma_3, \sigma_4$ aus der Permutation 1, 2, 3, 4 durch eine gerade oder ungerade Zahl von Vertauschungen zweier Indizes hervorgeht. Falls $(e_{\sigma_1 \ldots \sigma_4})$ ein Tensor ist, ist es nach dieser Definition ein Vektor; dass es ein Tensor ist, ~~sieht man in folgender Weise~~ kann unter Benutzung bekannter Determinantensätze leicht bewiesen werden.

§ 12.

Addition und Multiplikation der Tensoren.

Addition und Subtraktion von Tensoren.

Aus (34) geht hervor, dass man wieder die Komponenten eines Tensors erhält, wenn man die ~~gleichnamigen~~ entsprechenden ~~Tensoren~~ Komponenten zweier Tensoren desselben Ranges addiert oder subtrahiert. Man spricht in diesem Sinne von der Addition bezw. Subtraktion von Tensoren. Es lässt sich also schreiben

$$(T_{\sigma_1 \ldots \sigma_n}) \pm (U_{\sigma_1 \ldots \sigma_n}) = (T_{\sigma_1 \ldots \sigma_n} \pm U_{\sigma_1 \ldots \sigma_n}) \quad \ldots \quad (35)$$

Symmetrische Tensoren ~~geben dabei~~ liefern dabei als Summe bezw. Differenz wieder symmetrische Tensoren, Vektoren liefern wieder Vektoren.

Bemerkung. Wir sind imstande, zu jedem Tensor zweiten Ranges einen symmetrischen Tensor und einen Vektor zweiten Ranges zu bilden. Ist $(T_{\sigma\tau})$ der gegebene Tensor, so ist nach (33) $(T_{\tau\sigma})$ ebenfalls ein Tensor und daher

$(T_{\sigma\tau} + T_{\tau\sigma})$ ein symmetrischer Tensor zweiten Ranges

$(T_{\sigma\tau} - T_{\tau\sigma})$ ein Vektor zweiten Ranges.

~~Innere~~ Äussere Multiplikation von Tensoren

Aus zwei Tensoren mit den Komponenten $T_{\sigma_1 \ldots \sigma_m}$ bezw. $U_{\tau_1 \ldots \tau_n}$ lässt sich ein neuer Tensor $(m+n)$-ten Ranges bilden mit den Komponenten

$$T_{\sigma_1 \ldots \sigma_m} U_{\tau_1 \ldots \tau_n}$$

Den Tensor $(T_{\sigma_1 \ldots \sigma_m} U_{\tau_1 \ldots \tau_n})$ nennen wir das äussere Produkt der beiden gegebenen Tensoren. Denn es ist nach (34)

$$T'_{s_1 \ldots s_m} = \sum_{\sigma_1 \ldots \sigma_m} \alpha_{s_1 \sigma_1} \alpha_{s_2 \sigma_2} \cdots \alpha_{s_m \sigma_m} T_{\sigma_1 \ldots \sigma_m}$$

$$U'_{t_1 \ldots t_n} = \sum_{\tau_1 \ldots \tau_n} \alpha_{t_1 \tau_1} \alpha_{t_2 \tau_2} \cdots \alpha_{t_n \tau_n} U_{\tau_1 \ldots \tau_n},$$

woraus durch Multiplikation folgt

$$T'_{s_1 \ldots s_m} U'_{t_1 \ldots t_n} = \sum_{\sigma_1 \ldots \sigma_m \tau_1 \ldots \tau_n} \alpha_{s_1 \sigma_1} \cdots \alpha_{s_m \sigma_m} \alpha_{t_1 \tau_1} \cdots \alpha_{t_n \tau_n} T_{\sigma_1 \ldots \sigma_m} U_{\tau_1 \ldots \tau_n}.$$

Thus, the construction of a new tensor by taking the outer product occurs according to the schema:

$$(T_{\sigma_1 \ldots \sigma_m})(U_{\tau_1 \ldots \tau_n}) = (T_{\sigma_1 \ldots \sigma_m} U_{\tau_1 \ldots \tau_n}). \quad \ldots (35)$$

Remark: If we multiply all of the components of a tensor with one and the same number (scalar) that is invariant with respect to coordinate transformations, we obtain again, as follows from (34), a tensor of the same rank. One can view this multiplication as the taking of an outer product in the sense of (35) in that one considers the scalar as a tensor of rank zero; this conception of the scalar is expedient in any case.
Remark: If one forms the outer product of two symmetric tensors or two vectors, then the symmetry character or the vector character is not preserved. But one can again form symmetric tensors or vectors from the result by commuting the indices and adding or subtracting.
Remark. Equation (35) shows that the factors are interchangeable.

Inner products of Tensors

Let there be given a tensor of rank m $(T_{\sigma_1 \ldots \sigma_m})$ and a tensor of rank n $(U_{\tau_1 \ldots \tau_n})$, and let $m > n$. One can then show that

$$\sum_{\sigma_1 \ldots \sigma_n} U_{\sigma_1 \ldots \sigma_n} T_{\sigma_1 \ldots \sigma_m} = V_{\sigma_{n+1} \ldots \sigma_m} \quad \ldots (36)$$

is a tensor of rank $m - n$, which is called "the inner product" of the tensors $(U_{\tau_1 \ldots \tau_n})$ and $(T_{\sigma_1 \ldots \sigma_m})$. The proof that $V_{\sigma_{n+1} \ldots \sigma_m}$ is a tensor is arrived at in the following way. It follows from (34) that

$$U'_{s_1 \ldots s_n} = \sum_{\tau_1 \ldots \tau_n} \alpha_{s_1\tau_1} \alpha_{s_2\tau_2} \ldots \alpha_{s_n\tau_n} U_{\tau_1 \ldots \tau_n}$$

$$T'_{s_1 \ldots s_m} = \sum_{\sigma_1 \ldots \sigma_m} \alpha_{s_1\sigma_1} \alpha_{s_2\sigma_2} \ldots \alpha_{s_m\sigma_m} T_{\sigma_1 \ldots \sigma_m}.$$

From this it follows that

$$V'_{s_{n+1} \ldots s_m} = \sum_{s_1 \ldots s_n \tau_1 \ldots \tau_n \sigma_1 \ldots \sigma_m} \alpha_{s_1\tau_1} \ldots \alpha_{s_n\tau_n} \alpha_{s_1\sigma_1} \ldots \alpha_{s_m\sigma_m} U_{\tau_1 \ldots \tau_n} T_{\sigma_1 \ldots \sigma_m}.$$

But according to (16)

$\sum_{s_1} \alpha_{s_1\tau_1} \alpha_{s_1\sigma_1} = 1$ or $= 0$, depending on whether $\tau_1 = \sigma_1$ or $\tau_1 \neq \sigma_1$, etc.

Hence one obtains, in the light of (36),

$$V'_{s_{n+1} \ldots s_m} = \sum_{\sigma_{n+1} \ldots \sigma_m} \alpha_{s_{n+1}\sigma_{n+1}} \ldots \alpha_{s_m\sigma_m} V_{\sigma_{n+1} \ldots \sigma_m},$$

which equation, by virtue of (34), proves the assertion.
Let us write (36) in the form

$$(U_{\sigma_1 \ldots \sigma_n})(T_{\sigma_1 \ldots \sigma_m}) = (V_{\sigma_{n+1} \ldots \sigma_m}) \quad \ldots (36a)$$

Thus, we formulate the construction of the inner product the same way as that of the outer product. That we have here an inner product is expressed by the fact that those indices of the tensors that are involved in the inner multiplication are denoted in the two factors by the same letters.

Die Bildung eines neuen Tensors durch äussere Multiplikation geschieht also nach dem Schema:

$$(T_{\sigma_1\ldots\sigma_m})(U_{\tau_1\ldots\tau_n}) = (T_{\sigma_1\ldots\sigma_m}U_{\tau_1\ldots\tau_n}) \quad \ldots\ldots (35)$$

Bemerkung. Wenn wir alle Komponenten eines Tensors mit ein und derselben bezüglich ~~der Koord~~ Koordinatentransformationen invarianten Zahl (Skalar) multiplizieren, so erhalten wir, wie aus (34) hervorgeht wieder einen Tensor vom gleichen Range. Man kann diese Multiplikation als eine äussere im Sinne von (35) ansehen, indem man den Skalar als einen Tensor nullten Ranges betrachtet; diese Auffassung des Skalars ist überhaupt zweckmässig.

Bemerkung. Bildet man aus zwei symmetrischen Tensoren bezw aus zwei Vektoren das äussere Produkt, so bleibt der Charakter der Symmetrie bezw. der Vektorcharakter nicht gewahrt. Es lassen sich aber durch Kommutieren von Indizes und Addition bezw. Subtraktion aus dem Resultat wieder symmetrische ~~Vektoren~~ Tensoren bezw. Vektoren bilden.

Bemerkung. Gleichung (35) zeigt, dass die Faktoren vertauschbar sind. ~~Es lassen sich ferner äussere Produkte mehrerer Tensoren bilden, wobei Reihenfolge~~

Innere Multiplikation von Tensoren.

Es sei ein Tensor vom m-ten Range $(T_{\sigma_1\ldots\sigma_m})$ und ein Tensor vom n-ten Range $(U_{\tau_1\ldots\tau_n})$ gegeben, und es sei $m > n$. Dann lässt sich zeigen, dass

$$\sum_{\sigma_1\ldots\sigma_n} U_{\sigma_1\ldots\sigma_n} T_{\sigma_1\ldots\sigma_m} = V_{\sigma_{n+1}\ldots\sigma_m} \quad \ldots\ldots (36)$$

ein Tensor vom Range $m-n$ ist, den man das „innere Produkt" der ~~gegebenen~~ Tensoren $(U_{\tau_1\ldots\tau_n})$ und $(T_{\sigma_1\ldots\sigma_m})$ nennt. Der Beweis, dass $V_{\sigma_{n+1}\ldots\sigma_m}$ ein Tensor ist, ergibt sich folgendermassen. Aus (34) folgt

$$U'_{s_1\ldots s_n} = \sum_{\tau_1\ldots\tau_n} \alpha_{s_1\tau_1}\alpha_{s_2\tau_2}\ldots\alpha_{s_n\tau_n} U_{\tau_1\ldots\tau_n}$$

$$T'_{s_1\ldots s_m} = \sum_{\sigma_1\ldots\sigma_m} \alpha_{s_1\sigma_1}\alpha_{s_2\sigma_2}\ldots\alpha_{s_m\sigma_m} T_{\sigma_1\ldots\sigma_m}$$

Hieraus folgt

$$V'_{s_{n+1}\ldots s_m} = \sum_{s_1\ldots s_n \tau_1\ldots\tau_n \sigma_1\ldots\sigma_m} \alpha_{s_1\tau_1}\ldots\alpha_{s_n\tau_n}\alpha_{s_1\sigma_1}\ldots\alpha_{s_m\sigma_m} U_{\tau_1\ldots\tau_n} T_{\sigma_1\ldots\sigma_m}$$

Es ist aber nach (16)

$$\sum_{s_1} \alpha_{s_1\tau_1}\alpha_{s_1\sigma_1} = 1 \text{ oder } = 0,$$ je nachdem $\tau_1 = \sigma_1$ oder $\tau_1 \neq \sigma_1$, etc.

Deshalb erhält man mit Rücksicht auf (36)

$$V'_{s_{n+1}\ldots s_m} = \sum_{\sigma_{n+1}\ldots\sigma_m} \alpha_{s_{n+1}\sigma_{n+1}}\ldots\alpha_{s_m\sigma_m} V_{\sigma_{n+1}\ldots\sigma_m},$$

welche Gleichung gemäss (34) die Behauptung beweist.

Wir schreiben (36) in der Form:

$$(U_{\sigma_1\ldots\sigma_n})(T_{\sigma_1\ldots\sigma_m}) = (V_{\sigma_{n+1}\ldots\sigma_m}) \quad \ldots\ldots (36a)$$

Wir drücken also die innere Produktbildung ebenso aus wie die äussere Produktbildung. Dass es sich hier um innere Produktbildung handelt, drückt sich dadurch aus, dass ~~beide Tensoren n gleich bezeichnete Indizes haben.~~ diejenigen Indizes der Tensoren, auf welche sich die innere Multiplikation zu beziehen hat, in beiden Faktoren mit denselben Buchstaben bezeichnet sind.

Examples.
1) Inner product of two first-rank tensors (four-vectors)

$$(T_\mu)(U_\mu) = (V)$$
$$V = \sum_\mu T_\mu U_\mu .$$

The result is a zero-rank tensor, and thus, a scalar.
2) Inner product of second-rank tensor with a first-rank tensor

$$(T_{\mu\nu})(U_\nu) = (V_\mu)$$
$$V_\mu = \sum_\nu T_{\mu\nu} U_\nu .$$

The result is a four-vector. A second possibility, which yields the same result only in the case of symmetric tensors, is given by $\sum T_{\mu\nu} U_\mu$.
3) Inner product of two second-rank tensors:

$$(T_{\mu\nu})(U_{\mu\nu}) = (V)$$
$$V = \sum_{\mu\nu} T_{\mu\nu} U_{\mu\nu} .$$

The result is a scalar. An interesting special case of this is that in which the special tensor $(\delta_{\mu\nu})$ introduced in the previous § is substituted for $(U_{\mu\nu})$. In that case one obtains

$$(T_{\mu\nu})(\delta_{\mu\nu}) = (V)$$
$$V = \sum_{\mu\nu} T_{\mu\nu}\delta_{\mu\nu} = \sum_\mu T_{\mu\mu} .$$

Thus, the sum of the components of the main diagonal of a second-rank tensor is a scalar.[80] Another special case is that in which the tensors $(T_{\mu\nu})$ and $(U_{\mu\nu})$ are identical. One obtains

$$(T_{\mu\nu})(T_{\mu\nu}) = (V)$$
$$V = \sum T_{\mu\nu}^2 .$$

Thus, the sum of the squares of all of the components of a tensor is always a scalar; this theorem holds for tensors of arbitrary rank. The scalar so obtained can be called the "magnitude" of the tensor.

Mixed Multiplication

Both kinds of tensor multiplication presented so far are special cases of a more general type of multiplication that also yields from two tensors a third one. That is to say, the multiplication can be an inner product with respect to some tensor indices and an outer one with respect to the rest. We will give this most general multiplication rule without proof, since the proof is merely a combination of the two proofs we have already given.

The schema of this multiplication is the following:

$$\left.\begin{aligned} &(T_{s_1 \ldots s_l \sigma_1 \ldots \sigma_m})(U_{s_1 \ldots s_l \tau_1 \ldots \tau_n}) = (V_{\sigma_1 \ldots \sigma_m \tau_1 \ldots \tau_n}), \\ &\text{where} \\ &V_{\sigma_1 \ldots \sigma_m \tau_1 \ldots \tau_n} = \sum_{s_1 \ldots s_l} T_{s_1 \ldots s_l \sigma'_1 \ldots \sigma_m} U_{s_1 \ldots s_l \tau_1 \ldots \tau_n} \end{aligned}\right\} \quad (37)$$

Example. Mixed multiplication of two second-rank tensors

Beispiele.

1) Innere Multiplikation zweier Tensoren ersten Ranges (Vierervektoren)

$$(T_\mu)(U_\mu) = (V)$$

$$V = \sum_\mu T_\mu U_\mu$$

Das Resultat ist ein Tensor nullten Ranges, also ein Skalar.

2) Innere Multiplikation eines Tensors zweiten Ranges und eines Tensors ersten Ranges

$$(T_{\mu\nu})(U_\nu) = (V_\mu)$$

$$V_\mu = \sum_\nu T_{\mu\nu} U_\nu$$

~~3) Innere~~ Das Resultat ist ein Vierervektor. Eine zweite Möglichkeit, die nur im Falle symmetrischer Tensoren das gleiche Resultat liefert, ist durch $\sum T_{\mu\nu} U_\mu$ gegeben.

3) Innere Multiplikation zweier Tensoren zweiten Ranges

$$(T_{\mu\nu})(U_{\mu\nu}) = (V)$$

$$V = \sum_{\mu\nu} T_{\mu\nu} U_{\mu\nu}$$

Das Resultat ist ein Skalar. Ein interessanter Spezialfall davon ist der, dass $(U_{\mu\nu})$ durch den im vorigen § eingeführten speziellen Tensor $(\delta_{\mu\nu})$ ersetzt wird. Dann erhält man

$$(T_{\mu\nu})(\delta_{\mu\nu}) = (V)$$

$$V = \sum_{\mu\nu} T_{\mu\nu}\delta_{\mu\nu} = \sum_\mu T_{\mu\mu}.$$

Die Summe der Hauptdiagonal-~~Gl~~ Komponenten eines Tensors zweiten Ranges ist also ein Skalar. Ein anderer Spezialfall ist der, dass die Tensoren $(T_{\mu\nu})$ und $(U_{\mu\nu})$ einander gleich sind. Man erhält

$$(T_{\mu\nu})(T_{\mu\nu}) = (V)$$

$$V = \sum T_{\mu\nu}^2$$

Die Summe der Quadrate aller Komponenten eines Tensors ist also stets ein Skalar; dieser Satz gilt für Tensoren beliebigen Ranges. Man kann den so erhaltenen Skalar ~~auch~~ als den „Betrag" des Tensors bezeichnen.

<u>Gemischte Multiplikation.</u>

Die beiden bisher genannten Multiplikationsarten von Tensoren sind Spezialfälle einer allgemeineren Multiplikationsart, die ebenfalls aus zwei Tensoren einen dritten Tensor liefert. Die Multiplikation kann nämlich bezüglich gewisser Tensorindizes eine innere, bezüglich der übrigen eine äussere sein. Wir geben diese allgemeinste Multiplikationsregel ohne Beweis, da letzterer nur eine Kombination der beiden angegebenen Beweise ist.

Das Schema dieser Multiplikation ist folgendes

$$\left.\begin{aligned} &(T_{s_1\cdots s_l\sigma_1\cdots\sigma_m})(U_{s_1\cdots s_l\tau_1\cdots\tau_n}) = (V_{\sigma_1\cdots\sigma_m\tau_1\cdots\tau_n}),\\ &\text{wobei}\\ &V_{\sigma_1\cdots\sigma_m\tau_1\cdots\tau_n} = \sum_{s_1\cdots s_l} T_{s_1\cdots s_l\sigma_1\cdots\sigma_m} U_{s_1\cdots s_l\tau_1\cdots\tau_n} \end{aligned}\right\} \quad (37)$$

Beispiel. Gemischte Multiplikation zweier Tensoren zweiten Ranges

$$(T_{\mu\nu})(U_{\nu\rho}) = (V_{\mu\rho})$$
$$V_{\mu\rho} = \sum_{\nu} T_{\mu\nu} U_{\nu\rho}$$

Complement. We can obtain a new tensor from a tensor if we multiply the former in some way or other by the special fourth-rank vector (e_{iklm}). If this is an inner product, we call the result (disregarding a numerical factor) the *complement* of the original tensor. The following cases are of interest:

a) Complement of the four-vector

$$(T_i)(e_{iklm}) = (\overset{|}{V}_{klm})$$
$$V_{klm} = \sum_{i} T_i e_{iklm}.$$

Only *one* summand in the sum is different from 0, so that the components are equal to $\pm T_i$. If one again forms the complement of this, one obtains

$$\sum_{iklm} T_i\, e_{iklm}\, e_{jklm} = 6T_j,$$

which is again the original four-vector.

b) Complement of the six-vector

$$\left.\begin{aligned} \frac{1}{2}(\overset{|}{T}_{ik})\,(e_{iklm}) &= (\overset{|}{V}_{lm}) \\ V_{lm} = -V_{ml} &= \frac{1}{2}\sum T_{ik} e_{iklm} \end{aligned}\right\} \qquad (35)$$

Thus, the complement of a six-vector is again a six-vector, whose components, apart from the assignment to indices (coordinate axes), are the same as those of the original six-vector. We denote the complement of the six-vector (T_{ik}) <the associated six-vector> by (T_{ik}^*). The following schema shows the way in which the components are mapped

index combination ik	$\overset{|}{T}_{ik}$	$\overset{|}{T}{}^*_{ik}$
23	$\overset{|}{T}_{23}$	$\overset{|}{T}_{14}$
31	$\overset{|}{T}_{31}$	$\overset{|}{T}_{24}$
12	$\overset{|}{T}_{12}$	$\overset{|}{T}_{34}$
14	$\overset{|}{T}_{14}$	$\overset{|}{T}_{23}$
24	$\overset{|}{T}_{24}$	$\overset{|}{T}_{31}$
34	$\overset{|}{T}_{34}$	$\overset{|}{T}_{12}$

(35a)

If one forms a complement to the complement, one again arrives at the original six-vector. Thus, (T_{ik}) is also the complement of T_{ik}.[81]

c) Complement of the vector of the 3d rank

$$\frac{1}{6}(T_{ikl})(e_{iklm}) = (V_m)$$
$$V_m = \frac{1}{6}\sum_{ikl} T_{ikl}\, e_{iklm}.$$

The complement is a four-vector whose components are numerically identical with

~~$(T_{\mu\nu})(\mathfrak{V}_{\nu\varrho}) =$~~

$$(T_{\mu\nu})(U_{\nu\varrho}) = (V_{\mu\varrho})$$

$$V_{\mu\varrho} = \sum_{\nu} T_{\mu\nu} U_{\nu\varrho}$$

<u>Ergänzung</u>. Wir können aus einem Tensor einen neuen Tensor dadurch gewinnen, dass wir ihn auf irgend welche Art mit dem speziellen Vektor vierten Ranges (ϵ_{iklm}) multiplizieren. Ist diese Multiplikation eine (abgesehen von einem ~~Faktor~~ Zahlenfaktor) innere, so nennen wir das Resultat ~~der Multipli~~ die Ergänzung des ursprünglichen Tensors. Von Interesse sind folgende Fälle

a) Ergänzung des Vierervektors

$$(T_i)(\epsilon_{iklm}) = (\dot{V}_{klm})$$

$$V_{klm} = \sum_i T_i \epsilon_{iklm}$$

In der Summe ist nur <u>ein</u> Summand von 0 verschieden, sodass die Komponenten gleich $\pm T_i$ sind. Bildet man hievon wieder die Ergänzung, so erhält man

$$\sum_{ijklm} T_i \epsilon_{iklm} \epsilon_{jklm} = 6T_j,$$

also wieder den ursprünglichen Vierervektor.

b) Ergänzung des Sechservektors

$$\left.\begin{aligned} \tfrac{1}{2}(\dot{T}_{ik})(\epsilon_{iklm}) &= (\dot{V}_{lm}) \\ V_{lm} = -V_{ml} &= \tfrac{1}{2}\sum T_{ik}\epsilon_{iklm} \end{aligned}\right\} \quad (35)$$

Die Ergänzung des Sechservektors ist also wieder ein Sechservektor, ~~die Komponenten~~ dessen Komponenten abgesehen von der Zuordnung zu den Indizes (Koordinatenachsen) dieselben sind wie diejenigen des ursprünglichen Sechservektors. ~~Den dem~~ Die Ergänzung des Sechservektors $(\dot{T}_{ik})$ ~~zugeordneten Sechservektor~~ nennen wir $(\dot{T}^{*}_{ik})$. Die Art der Zuordnung der Komponenten zeigt folgendes Schema

Indizes Kombination ik	$\dot{T}_{ik}$	$\dot{T}^{*}_{ik}$
23	$\dot{T}_{23}$	$\dot{T}_{14}$
31	$\dot{T}_{31}$	$\dot{T}_{24}$
12	$\dot{T}_{12}$	$\dot{T}_{34}$
14	$\dot{T}_{14}$	$\dot{T}_{23}$
24	$\dot{T}_{24}$	$\dot{T}_{31}$
34	$\dot{T}_{34}$	$\dot{T}_{12}$

(35a)

Bildet man zu der Ergänzung nochmals die Ergänzung, so gelangt man wieder zu dem ursprünglichen Sechservektor. $(\dot{T}_{ik})$ ist also auch die Ergänzung von T^{*}_{ik}.

c) Ergänzung des Vektors 3. Ranges

$$\tfrac{1}{6}(\dot{T}_{ikl})(\epsilon_{iklm}) = (V_m)$$

$$V_m = \tfrac{1}{6}\sum_{ikl} T_{ikl}\epsilon_{iklm}$$

Die Ergänzung ist ein Vierervektor, dessen Komponenten numerisch gleich sind

the components of the vector of the third rank.

d) Complement of the vector of the fourth rank

$$\frac{1}{24}(T_{iklm})(e_{iklm}) = V$$

$$V = \frac{1}{24}\sum_{iklm} T_{iklm}\, e_{iklm}.$$

The complement is a scalar that is numerically identical with the common magnitude of the components of the vector $(\overset{\text{I}}{T}_{iklm})$.

Remark: Instead of the vectors of the 3d and 4th rank, one usually employs their complements.

§18. Differential Covariants

As the inverse of the system of equations (31) one obtains

$$x_\nu = \sum_\mu \alpha_{\mu\nu} x'_\mu$$

and from this

$$\frac{\partial}{\partial x'_\mu} = \sum_\nu \alpha_{\mu\nu} \frac{\partial}{\partial x_\nu}. \qquad \ldots (36)$$

Thus, as a comparison of (36) with (32) shows, the operators $\frac{\partial}{\partial x_1} \cdots \frac{\partial}{\partial x_4}$ have the same transformation properties as those of the components of a four-vector. We can therefore combine these four operators symbolically into a four-vector $(\frac{\partial}{\partial x_\nu})$, after the fashion of the theory of invariants, noting that the way in which one combines this symbolic vector with the structure to be differentiated is completely analogous to that followed in multiplication. Hence there are two ways of forming a new tensor from a tensor by differentiation, namely:

1) Extension ("outer differentiation") according to the schema [82]

$$(\frac{\partial}{\partial x_\tau})(T_{\sigma_1 \ldots \sigma_n}) = \left(\frac{\partial T_{\sigma_1 \ldots \sigma_n}}{\partial x_\tau}\right); \qquad \ldots (37)$$

in this way one obtains an $(n+1)$-rank tensor from an n-rank tensor.

2) Divergence (inner differentiation) according to the schema

$$\left(\frac{\partial}{\partial x_{\sigma_\nu}}\right)(T_{\sigma_1 \ldots \sigma_n}) = \sum_{\sigma_\nu} \frac{\partial}{\partial x_{\sigma_\nu}}(T_{\sigma_1 \ldots \sigma_{\nu-1}\sigma_{\nu+1} \ldots \sigma_n}), \qquad \ldots (38)$$

where σ_ν denotes one of the indices $\sigma_1 \ldots \sigma_\nu$. Thus, there are n divergences of a tensor of the nth rank. They are identical to one another in the case of the symmetric tensor, and differ in the case of the vector only in sign.

Examples.

Extension of the scalar

$$(\frac{\partial}{\partial x_\nu})T = (\frac{\partial T}{\partial x_\nu}).$$

The result of the extension is a four-vector

Extension of the four-vector

$$(\frac{\partial}{\partial x_\nu})(T_\mu) = \left(\frac{\partial T_\mu}{\partial x_\nu}\right)$$

The result is a tensor of the second rank, which leads to the construction of the symmetric tensor of the second rank

$$\left(\frac{\partial T_\mu}{\partial x_\nu} + \frac{\partial T_\nu}{\partial x_\mu}\right)$$

der Komponenten des Vektors dritten Ranges

d) Ergänzung des Vektors 4. Ranges

$$\frac{1}{24}(\overset{1}{T}_{iklm})(e_{iklm}) = V$$

$$V = \frac{1}{24}\sum_{iklm} T_{iklm} e_{iklm}$$

Die Ergänzung ist ein Skalar, der numerisch gleich ist dem gemeinsamen Betrage der Komponenten des Vektors $(\overset{1}{T}_{iklm})$

Bemerkung. Statt der Vektoren 3. und 4. Ranges werden gewöhnlich deren Ergänzungen verwendet.

§18. Differential-Invarianten.

Als Umkehrung des Gleichungssystems (31) erhält man

$$x_\nu = \sum_\mu \alpha_{\mu\nu} x'_\mu$$

und hieraus

$$\frac{\partial}{\partial x'_\mu} = \sum_\nu \alpha_{\mu\nu} \frac{\partial}{\partial x_\nu} \quad \ldots\ldots (36)$$

Es sind also, wie ein Vergleich von (36) mit (32) lehrt, die Transformationseigenschaften der Operatoren $\frac{\partial}{\partial x_1} \cdots \frac{\partial}{\partial x_4}$ dieselben wie diejenigen der Komponenten eines Vierervektors. Wir können also invariantentheoretisch diese vier Operatoren symbolisch zu einem Vierervektor $\left(\frac{\partial}{\partial x_\nu}\right)$ zusammenfassen, indem wir hinzufügen, dass die Art der Zusammenfügung dieses symbolischen Vektors mit dem zu differenzierenden Gebilde derjenigen bei der Multiplikation vollkommen gleichartig ist. Es gibt daher zwei Arten, um durch Differenzieren aus einem Tensor einen neuen zu bilden, nämlich:

1) die Erweiterung („äussere Differenziation") nach dem Schema

$$\left(\frac{\partial}{\partial x_\tau}\right)\left(T_{\sigma_1 \ldots \sigma_n}\right) = \left(\frac{\partial T_{\sigma_1 \ldots \sigma_n}}{\partial x_\tau}\right); \quad \ldots\ldots (37)$$

man erhält so aus einem Tensor vom n-ten Range einen vom $(n+1)$ten Range

2) die Divergenz (innere Differentiation) nach dem Schema

$$\left(\frac{\partial}{\partial x_{\sigma_\nu}}\right)\left(T_{\sigma_1 \ldots \sigma_n}\right) = \sum_{\sigma_\nu} \frac{\partial}{\partial x_{\sigma_\nu}}\left(T_{\sigma_1 \ldots \sigma_{\nu-1}\sigma_{\nu+1} \ldots \sigma_n}\right), \ldots (38)$$

wobei σ_ν einen der Indizes $\sigma_1 \cdots \sigma_n$ bedeutet. Es gibt also n Divergenzen eines Tensors n-ten Ranges. Diese sind im Falle des symmetrischen Tensors einander gleich und unterscheiden sich im Falle des Vektors nur durch das Vorzeichen.

Beispiele.

Erweiterung des Skalars

$$\left(\frac{\partial}{\partial x_\nu}\right) T = \left(\frac{\partial T}{\partial x_\nu}\right).$$

Resultat der Erweiterung ist ein Vierervektor

Erweiterung des Vierervektors

$$\left(\frac{\partial}{\partial x_\nu}\right)\left(T_\mu\right) = \left(\frac{\partial T_\mu}{\partial x_\nu}\right)$$

Resultat ist ein Tensor zweiten Ranges, der zur Bildung des symmetrischen Tensors zweiten Ranges

$$\left(\frac{\partial T_\mu}{\partial x_\nu} + \frac{\partial T_\nu}{\partial x_\mu}\right)$$

and the six-vector

$$(\overset{|}{U}_{\mu\nu}) = \left(\frac{\partial T_\mu}{\partial x_\nu} - \frac{\partial T_\nu}{\partial x_\mu}\right) \quad \ldots (39)$$

Divergence of the four-vector

$$(\frac{\partial}{\partial x_\nu})(T_\nu) = \sum_\nu \frac{\partial T_\nu}{\partial x_\nu}$$

The result is a scalar.

Divergence of a tensor of the second rank

$$(\frac{\partial}{\partial x_\nu})(T_{\mu\nu}) = \left(\sum \frac{\partial T_{\mu\nu}}{\partial x_\nu}\right)$$

The result is a four-vector.

Of interest is also the tensor that is produced when one forms the divergence of the extension of a tensor, where the construction of the divergence is carried out with respect to the index that comes in as a result of the extension. One thus obtains [83]

$$(\frac{\partial}{\partial x_\tau})\{(\frac{\partial}{\partial x_\tau})(T_{\sigma_1 \ldots \sigma_n})\} = (\frac{\partial}{\partial x_\tau})\left(\frac{\partial T_{\sigma_1 \ldots \sigma_n}}{\partial x_\tau}\right) =$$

$$\left(\sum_\tau \frac{\partial^2 T_{\sigma_1 \ldots \sigma_n}}{\partial x_\tau^2}\right) = \left(\Box\, T_{\sigma_1 \ldots \sigma_n}\right) \quad \ldots (40)$$

Thus, this operation, which is a generalization of the Laplacian operator, produces from a tensor of the *n*th rank again a tensor of the *n*th rank, with the symmetry or vector character preserved.

und des Sechservektors

$$\left(\overset{1}{\mathfrak{U}}_{\mu\nu}\right) = \left(\frac{\partial T_\mu}{\partial x_\nu} - \frac{\partial T_\nu}{\partial x_\mu}\right) \cdots\cdots (39)$$

Veranlassung gibt.

Divergenz des Vierervektors

$$\left(\frac{\partial}{\partial x_\nu}\right)\left(T_\nu\right) = \sum_\nu \frac{\partial T_\nu}{\partial x_\nu}$$

Das Resultat ist ein Skalar.

Divergenz eines Tensors zweiten Ranges

$$\left(\frac{\partial}{\partial x_\nu}\right)\left(T_{\mu\nu}\right) = \left(\sum \frac{\partial T_{\mu\nu}}{\partial x_\nu}\right)$$

Das Resultat ist ein Vierervektor.

Von Interesse ist ferner der Tensor, der entsteht, wenn man die Divergenz der Erweiterung ~~bild~~ eines Tensors bildet, wobei die Divergenzbildung nach demjenigen Index vorgenommen wird, der durch die Erweiterung hinzukommt. Man erhält so

$$\left(\frac{\partial}{\partial x_\tau}\right)\left\{\left(\frac{\partial}{\partial x_\tau}\right)\left(T_{\sigma_1 \ldots \sigma_n}\right)\right\} = \left(\frac{\partial}{\partial x_\tau}\right)\left(\frac{\partial T_{\sigma_1 \ldots \sigma_n}}{\partial x_\tau}\right) = \left(\sum_\tau \frac{\partial^2 T_{\sigma_1 \ldots \sigma_n}}{\partial x_\tau^2}\right) = \left(\Box T_{\sigma_1 \ldots \sigma_n}\right) \cdot\cdot (40)$$

~~Es ist~~ Diese Operation, welche eine Verallgemeinerung der Laplace'schen Operators ist, liefert also aus einem Tensor n-ten Ranges wieder einen Tensor n-ten Ranges, und zwar unter Erhaltung des Symmetrie- bezw. Vektor-Charakters.

SECTION 4

ELECTRODYNAMICS OF MOVING BODIES [84]

§18. The Fundamental Electromagnetic Equations for Empty Space

First we will consider the case of electrodynamics of the vacuum, as it was treated in §1 and §2, from the standpoint of Minkowski's theory of covariants. For this purpose we will now write equations (I) from §1 in a form that makes their covariance evident. Let $(\mathfrak{F}_{\mu\nu})$ be a six-vector with the components

$$\left.\begin{matrix} \mathfrak{F}_{23} = \mathfrak{h}_x & \mathfrak{F}_{31} = \mathfrak{h}_y & \mathfrak{F}_{12} = \mathfrak{h}_z \\ \mathfrak{F}_{14} = -i\mathfrak{e}_x & \mathfrak{F}_{24} = -i\mathfrak{e}_y & \mathfrak{F}_{34} = -i\mathfrak{e}_z \end{matrix}\right\} \quad (41)$$

and $(\overset{*}{\mathfrak{F}}_{\mu\nu})$ an associated dual six-vector (complement) with the components

$$\left.\begin{matrix} \mathfrak{F}^*_{23} = -i\mathfrak{e}_x & \mathfrak{F}^*_{31} = -i\mathfrak{e}_y & \mathfrak{F}^*_{12} = -i\mathfrak{e}_z \\ \mathfrak{F}^*_{14} = \mathfrak{h}_x & \mathfrak{F}^*_{24} = \mathfrak{h}_y & \mathfrak{F}^*_{34} = \mathfrak{h}_z \end{matrix}\right\} \quad (41a)$$

Furthermore, let $(\mathfrak{J}_\mu)$ be a four-vector with the components

$$\mathfrak{J}_1 = \frac{\mathfrak{q}_x}{c}\rho \qquad \mathfrak{J}_2 = \frac{\mathfrak{q}_y}{c}\rho \qquad \mathfrak{J}_3 = \frac{\mathfrak{q}_z}{c}\rho \qquad \mathfrak{J}_4 = i\rho. \quad (42)$$

Then equations (I) can be written in the form

$$\left(\frac{\partial}{\partial x_\nu}\right)(\mathfrak{F}_{\mu\nu}) = (\mathfrak{J}_\mu) \qquad \left(\frac{\partial}{\partial x_\nu}\right)(\mathfrak{F}^*_{\mu\nu}) = 0 \quad \ldots (43)$$

The fact that we were able to put equations (I) into this form tells us not only that these equations are covariant, i.e., that they transform to equations of the same form when a reference system moving uniformly relative to the original system is introduced, but at the same time it teaches us how the components of the electromagnetic field and of the electrical (convection) current are to be transformed in the case of this change. The transformation law is obtained from (33) and (32) by using (41) and (42). If one carries out the special Lorentz transformation (§15), which is given by the schema

	1	2	3	4
1′	$\frac{1}{\sqrt{1-\beta^2}}$	0	0	$\frac{-i\beta}{\sqrt{1-\beta^2}}$
2′	0	1	0	0
3′	0	0	1	0
4′	$+\frac{i\beta}{\sqrt{1-\beta^2}}$	0	0	$\frac{1}{\sqrt{1-\beta^2}}$

$$\beta = \frac{v}{c}$$

4. Abschnitt.

Elektrodynamik bewegter Körper.

§18. Die elektromagnetischen Grundgleichungen für den leeren Raum.

Wir wollen zunächst den Fall der Elektrodynamik des Vakuums, so wie er in §1 und §2 behandelt ist, vom Standpunkt der Minkowskischen Kovarianten-Theorie betrachten. Zu diesem Zweck schreiben ~~wir~~ zunächst die Gleichungen (I) des §1 in solcher Form, dass deren Kovarianz ~~klar~~ erkennbar ist. Es sei $(\mathfrak{F}_{\mu\nu})$ ein Sechservektor mit den Komponenten

$$\left.\begin{array}{lll} \mathfrak{F}_{23} = \mathfrak{h}_x & \mathfrak{F}_{31} = \mathfrak{h}_y & \mathfrak{F}_{12} = \mathfrak{h}_z \\ \mathfrak{F}_{14} = -i\mathfrak{e}_x & \mathfrak{F}_{24} = -i\mathfrak{e}_y & \mathfrak{F}_{34} = -i\mathfrak{e}_z \end{array}\right\} \quad (41)$$

und $(\mathfrak{F}^*_{\mu\nu})$ der zugehörige duale (Ergänzung) Sechservektor ~~mit~~ den Komponenten

$$\left.\begin{array}{lll} \mathfrak{F}^*_{23} = -i\mathfrak{e}_x & \mathfrak{F}^*_{31} = -i\mathfrak{e}_y & \mathfrak{F}^*_{12} = -i\mathfrak{e}_z \\ \mathfrak{F}^*_{14} = \mathfrak{h}_x & \mathfrak{F}^*_{24} = \mathfrak{h}_y & \mathfrak{F}^*_{34} = \mathfrak{h}_z \end{array}\right\} \quad (41a)$$

Ferner sei $(\mathfrak{J}_\mu)$ ein Vierervektor mit den Komponenten

$$\mathfrak{J}_1 = \frac{\mathfrak{q}_x}{c}\varrho \quad \mathfrak{J}_2 = \frac{\mathfrak{q}_y}{c}\varrho \quad \mathfrak{J}_3 = \frac{\mathfrak{q}_z}{c}\varrho \quad \mathfrak{J}_4 = i\varrho. \quad (42)$$

Dann lassen sich die Gleichungen (I) in der Form schreiben:

~~$\frac{\partial \mathfrak{F}_{\mu\nu}}{\partial x_\nu}$~~ ~~$\sum \frac{\partial \mathfrak{F}_{\mu\nu}}{\partial x_\nu}$~~

$$\left(\frac{\partial}{\partial x_\nu}\right)(\mathfrak{F}_{\mu\nu}) = (\mathfrak{J}_\mu) \qquad \left(\frac{\partial}{\partial x_\nu}\right)(\mathfrak{F}^*_{\mu\nu}) = 0 \quad \ldots (43)$$

Damit, dass wir die Gleichungen (I) in diese Form bringen konnten, erfahren wir nicht nur, dass diese Gleichungen kovariant sind d.h. dass sie bei Einführung eines relativ zum ursprünglichen gleichförmig bewegten Bezugssystems in Gleichungen von derselben Form übergehen; sondern wir erfahren hierbei gleichzeitig, wie die Komponenten des ~~elektrischen und magnetischen~~ elektromagnetischen Feldes und des elektrischen (Konvektions-) Stromes ~~bei~~ bei dieser Änderung zu transformieren sind. Das Transformationsgesetz wird ~~mit Rücksicht auf~~ unter Benutzung von (41) und (42) aus (33) und (32) erhalten. Bei Ausführung der speziellen Lorentz-Transformation (§15), die durch das Transformationsschema

	1	2	3	4
1'	$\frac{1}{\sqrt{1-\beta^2}}$	0	0	$\frac{-i\beta}{\sqrt{1-\beta^2}}$
2'	0	1	0	0
3'	0	0	1	0
4'	$\frac{+i\beta}{\sqrt{1-\beta^2}}$	0	0	$\frac{1}{\sqrt{1-\beta^2}}$

$$\beta = \frac{v}{c}$$

one arrives at the transformation equations (21), (22), and (23). The circumstance that we were able to obtain the transformation equations for velocity from (42) assures us that we were right in assuming that those four quantities form a four-vector.

Let us note, by the way, that the following quantities are invariants of the Lorentz transformations

$$\frac{1}{2}(\mathfrak{F}_{\mu\nu})(\mathfrak{F}_{\mu\nu}) = \mathfrak{h}^2 - \mathfrak{e}^2 \qquad \ldots (44)$$

$$\frac{1}{2}i(\mathfrak{F}_{\mu\nu})(\mathfrak{F}^*{}_{\mu\nu}) = (\mathfrak{e}\mathfrak{h}) \qquad \ldots (45)$$

$$-(\mathfrak{J}_\nu)(\mathfrak{J}_\nu) = \rho^2\left(1 - \frac{q^2}{c^2}\right) = \rho_0^2 \qquad \ldots (46)$$

where ρ_0 is the electrical density ρ for a reference system for which $q = 0$, that is, for a reference system that is at rest relative to the electrical mass.

We shall now also bring the conservation laws for momentum and energy into a four-dimensional, covariant form. To that end we form the components of the four-vector

$$(\mathfrak{F}_{\nu\mu})(\mathfrak{J}_\nu) = (K_\mu).$$

We obtain

$$\left.\begin{aligned} K_1 &= \rho\,(\mathfrak{e}_x + \frac{\mathfrak{q}_y}{c}\mathfrak{h}_z - \frac{\mathfrak{q}_z}{c}\mathfrak{h}_y) \\ K_2 &= \rho\,(\mathfrak{e}_y + \frac{\mathfrak{q}_z}{c}\mathfrak{h}_x - \frac{\mathfrak{q}_x}{c}\mathfrak{h}_z) \\ K_3 &= \rho\,(\mathfrak{e}_z + \frac{\mathfrak{q}_x}{c}\mathfrak{h}_y - \frac{\mathfrak{q}_y}{c}\mathfrak{h}_x) \\ K_4 &= \frac{i}{c}\rho\,(\mathfrak{q}_x\mathfrak{e}_x + \mathfrak{q}_y\mathfrak{e}_y + \mathfrak{q}_z\mathfrak{e}_z) \end{aligned}\right\} \qquad (47)$$

A comparison of these equations with the results of §2 shows that the first three components of (K_μ) are identical to the components of the Lorentz force exerted on the electricity per unit volume $(\mathfrak{f}_\nu)$, while the fourth component is identical to the energy release of the electromagnetic field per unit volume and unit time, multiplied by $\frac{i}{c}$. With the help of the equations, we can also write the four-vector (K_μ) in the form

gegeben ist, gelangt man zu den Transformationsgleichungen (21), (22) und (23). In dem Umstande, dass wir die Transformationsgleichungen der Geschwindigkeit ~~durch~~ aus (42) erhalten konnten, liegt eine Gewähr dafür, dass wir mit Recht annahmen, dass jene vier Grössen einen Vierervektor bilden. ~~Dies lässt sich übrigens auch leicht direkt erweisen.~~

Wir bemerken nebenbei, dass folgende Grössen Invariante der Lorentz-Transformationen sind

$$\frac{1}{2}(\mathfrak{F}_{\mu\nu})(\mathfrak{F}_{\mu\nu}) = \mathfrak{f}^2 - \mathfrak{n}^2 \quad \ldots (44)$$

$$\frac{1}{2} i (\mathfrak{F}_{\mu\nu})(\mathfrak{F}^{*}_{\mu\nu}) = (\mathfrak{n}\mathfrak{f}) \quad \ldots (45)$$

$$-(\mathfrak{J}_\nu)(\mathfrak{J}_\nu) = \rho^2\left(1 - \frac{q^2}{c^2}\right) = \rho_0^2 \quad \ldots (46)$$

ρ_0 ist hiebei die elektrische Dichte ρ für ein solches Bezugssystem, für welches $q = 0$ ist, also für ein ~~mit~~ relativ der elektrischen Masse ruhendes Bezugssystem.

~~Wir wollen nun auch die Erhaltungssätze der Energie des Impulses und der Energie in eine vierdimensional kovariante Form bringen. Bezeichnen wir mit $\mathfrak{k}$ die pro Volumeneinheit auf die Elektrizität vom Felde ausgeübte ponderomotorische Kraft, mit η die vom Felde pro Zeiteinh Zeit- und Volumeneinheit auf die elektrischen Mengen abgegebene Energie, so kann man (6) und (5) in der Form schreiben:~~

$$\mathfrak{k}_x = \rho\left(\mathfrak{n}_x + \frac{q_y}{c}\mathfrak{f}_z - \frac{q_z}{c}\mathfrak{f}_y\right) = -\frac{\partial p_{xx}}{\partial x} - \frac{\partial p_{xy}}{\partial y} - \frac{\partial p_{xz}}{\partial z} - \frac{1}{c^2}\frac{\partial f_x}{\partial t}$$

- -

- -

$$-i\frac{\eta}{c} = \mathfrak{k} - i\rho\left(\frac{q_x}{c}\mathfrak{n}_x + \frac{q_y}{c}\mathfrak{n}_y + \frac{q_z}{c}\mathfrak{n}_z\right) =$$

Wir bilden zu diesem Zweck die Komponenten des Vierervektors

$$(\mathfrak{F}_{\mu\nu})(\mathfrak{J}_\nu) = (K_\mu).$$

Wir erhalten

$$\left.\begin{aligned}
K_1 &= \rho\left(\mathfrak{n}_x + \frac{q_y}{c}\mathfrak{f}_z - \frac{q_z}{c}\mathfrak{f}_y\right) \\
K_2 &= \rho\left(\mathfrak{n}_y + \frac{q_z}{c}\mathfrak{f}_x - \frac{q_x}{c}\mathfrak{f}_z\right) \\
K_3 &= \rho\left(\mathfrak{n}_z + \frac{q_x}{c}\mathfrak{f}_y - \frac{q_y}{c}\mathfrak{f}_x\right) \\
K_4 &= \frac{i}{c}\rho\left(q_x\mathfrak{n}_x + q_y\mathfrak{n}_y + q_z\mathfrak{n}_z\right)
\end{aligned}\right\} \quad (47)$$

Ein Vergleich dieser Gleichungen mit den Ergebnissen des §2 zeigt, dass die ersten drei Komponenten von (K_μ) denen der auf die Elektrizität pro Volumeneinheit ausgeübten Lorentz-Kraft $(\mathfrak{k}_\nu)$ gleich sind, während die vierte Komponente der mit $\frac{i}{c}$ multiplizierten Energieabgabe des elektromagnetischen Feldes pro Volumen- und Zeiteinheit gleich ist. Vermittelst der Gleichungen können wir den Vierervektor (K_μ) noch in der Form

$$\left.\begin{array}{l} K_1 = -\dfrac{\partial p_{xx}}{\partial x_1} - \dfrac{\partial p_{xy}}{\partial x_2} - \dfrac{\partial p_{xz}}{\partial x_3} - \dfrac{\partial \frac{i}{c}\mathbf{s}_x}{\partial x_4} \\ - - - - - - - - - - - - - - - - \\ - - - - - - - - - - - - - - - - \\ K_4 = -\dfrac{\partial \frac{i}{c}\mathbf{s}_x}{\partial x_1} - \dfrac{\partial \frac{i}{c}\mathbf{s}_y}{\partial x_2} - \dfrac{\partial \frac{i}{c}\mathbf{s}_z}{\partial x_3} - \dfrac{\partial (-w)}{\partial x_4} \end{array}\right\} \qquad (48)$$

or in the abbreviated form

$$(K_\mu) = -(\frac{\partial}{\partial x_\nu})\,(T_{\mu\nu}), \qquad \ldots (48a)$$

where we have set

$$(T_{\mu\nu}) = \frac{1}{2}\{(\mathfrak{F}_{\mu\sigma})(\mathfrak{F}_{\nu\sigma}) - (\mathfrak{F}^*{}_{\mu\sigma})(\mathfrak{F}^*{}_{\nu\sigma})\} \qquad \ldots (48b)$$

Thus, the Maxwell stresses, the components of the momentum density multiplied by ic, the components of the vector of the energy flow multiplied by $\frac{i}{c}$, and the component of the energy density, taken with a negative sign, form a symmetric tensor whose divergence, taken with a negative sign, is identical with the four-vector (K_μ). The transformation properties of this symmetric tensor follow from (33).

$$\left.\begin{array}{l} K_1 = -\frac{\partial p_{xx}}{\partial x_1} - \frac{\partial p_{xy}}{\partial x_2} - \frac{\partial p_{xz}}{\partial x_3} - \frac{\partial \frac{i}{c} f_x}{\partial x_4} \\ - - - - - - - - - - - - - \\ - - - - - - - - - - - - - \\ K_4 = -\frac{\partial \frac{i}{c} f_x}{\partial x_1} - \frac{\partial \frac{i}{c} f_y}{\partial x_2} - \frac{\partial \frac{i}{c} f_z}{\partial x_3} - \frac{\partial (-W)}{\partial x_4} \end{array}\right\} \quad (48)$$

oder kürzer in der Form

~~$(K_\mu) = -\left(\frac{\partial}{\partial x_\nu}\right)\left(\frac{\partial T_{\mu\nu}}{}\right)$~~

$$(K_\mu) = -\left(\frac{\partial}{\partial x_\nu}\right)(T_{\mu\nu}), \quad \ldots \ldots (48a)$$

wobei

$$(T_{\mu\nu}) = \frac{1}{2}\left\{(\mathfrak{F}_{\mu\sigma})(\mathfrak{F}_{\nu\sigma}) - (\mathfrak{F}^{*}_{\mu\sigma})(\mathfrak{F}^{*}_{\nu\sigma})\right\} \quad \ldots \ldots (48b)$$

gesetzt ist. Die Maxwell'schen Spannungen, die mit $i\,c$ multiplizierten Komponenten der Impulsdichte, die mit $\frac{i}{c}$ multiplizierten Komponenten des Vektors der Energieströmung und die negativ genommene Komponente der Energiedichte bilden also einen symmetrischen Tensor, dessen negativ genommene Divergenz gleich dem Vierervektor (K_μ) ist. Die Transformationseigenschaften dieses symmetrischen Tensors gehen aus (33) hervor

§19. Electrodynamics of Moving (Isotropic) Ponderable Bodies [85]

In the treatment of this subject I diverge from Minkowski, Abraham, and Laue's way of treating it, and use, instead, a more intuitive conception that follows more closely H. A. Lorentz's way of treating the subject, while I do employ the auxiliary mathematical tools created by Minkowski.[86]

As regards the matter, its motion is given first, i.e., the velocity $\mathfrak{q}$, to which the four-vector

$$(G_\mu) = \left(\frac{dx_\mu}{\sqrt{-\sum dx_\sigma^2}}\right) \qquad \ldots (49)$$

with the components

$$\left.\begin{aligned} G_1 &= \frac{\mathfrak{q}_x}{\sqrt{c^2-q^2}} \\ &- - - - - - - \\ G_4 &= \frac{ic}{\sqrt{c^2-q^2}} \end{aligned}\right\} \qquad (49a)$$

corresponds. Furthermore, there exists in the space in question a vacuum field characterized by the field vectors $\mathfrak{h}$ and $\mathfrak{e}$ and, four-dimensionally, by the six-vectors $(\mathfrak{F}_{\mu\nu})$ and $(\mathfrak{F}^*_{\mu\nu})$. From these and the four-vector of velocity we can form two four-vectors $(\mathfrak{F}_\mu)$ and $(\mathfrak{F}^*_\mu)$, which we call, respectively, the four-vectors of the electromotive and the magnetomotive force $(\mathfrak{F}_\mu^{(e)})$ and $(\mathfrak{F}_\mu^{m})$; let them be defined by the equations

$$\left.\begin{aligned} (\mathfrak{F}_\mu^{(e)}) &= (\mathfrak{F}_{\mu\nu})\,(G_\nu) \\ \mathfrak{F}_1^{(e)} &= \frac{1}{\sqrt{1-\frac{q^2}{c^2}}}\mathfrak{e}_x + \frac{\mathfrak{q}_y}{c}\mathfrak{h}_z - \frac{\mathfrak{q}_z}{c}\mathfrak{h}_y \\ &- - - - - - - - - - - - - - - \\ \mathfrak{F}_4^{(e)} &= \frac{i}{\sqrt{1-\frac{q^2}{c^2}}}\frac{\mathfrak{q}_x}{c}\mathfrak{e}_x + \frac{\mathfrak{q}_y}{c}\mathfrak{e}_y + \frac{\mathfrak{q}_z}{c}\mathfrak{e}_z \end{aligned}\right\} \qquad (50)$$

$$\left.\begin{aligned} (\mathfrak{F}_\mu^{(m)}) &= -i\,(\mathfrak{F}^*_{\mu\nu})\,(G_\nu) \\ \mathfrak{F}_1^{(m)} &= \frac{1}{\sqrt{1-\frac{q^2}{c^2}}}\mathfrak{h}_x - \frac{\mathfrak{q}_y}{c}\mathfrak{e}_z + \frac{\mathfrak{q}_z}{c}\mathfrak{e}_y \\ &- - - - - - - - - - - - - - - \\ \mathfrak{F}_4^{(m)} &= \frac{i}{\sqrt{1-\frac{q^2}{c^2}}}\frac{\mathfrak{q}_x}{c}\mathfrak{h}_x + \frac{\mathfrak{q}_y}{c}\mathfrak{h}_y + \frac{\mathfrak{q}_z}{c}\mathfrak{h}_z \end{aligned}\right\} \qquad (51)$$

Now it remains for us to define the electromagnetic state of ponderable matter in such a way that only four-dimensional covariant structures will be involved. Again, we proceed in such a way as to keep strictly separate from each other those things that possess an independent existence from a physical point of view.

a. *Electrical Polarization*

In media at rest, electrical polarization is represented by the vector $\mathfrak{p}$. We will let a four-vector $(\mathfrak{P}_\mu)$ correspond to it, with the additional stipulation that in the case of rest we should have (comoving system)

$$\mathfrak{P}_1 = \mathfrak{p}_x, \quad \mathfrak{P}_2 = \mathfrak{p}_y, \quad \mathfrak{P}_3 = \mathfrak{p}_z, \quad \mathfrak{P}_4 = 0$$

For the comoving system, and thus, in general, this four-vector satisfies the condition

$$(\mathfrak{P}_\mu)\,(G_\mu) = 0 \qquad \ldots (52)$$

In the light of this equation, and extending the meaning of the symbols $\mathfrak{p}_x$, we can set for arbitrarily moving bodies

§ 19. Elektrodynamik bewegter (isotroper) ponderabler Körper.

Bei der Behandlung dieses Gegenstandes weiche ich von der Behandlungsweise Minkowskis, Abrahams und Laues ab, ~~weil ich~~ und bediene mich einer anschaulicheren Begriffsbildung, welche mehr ~~im Sinne der~~ an die H. A. Lorentz'sche Behandlungsweise des Gegenstandes anschliesst, indem ich ~~mich~~ jedoch die von Minkowski geschaffenen mathematischen Hilfsmittel ~~bediene~~ verwende.

Von der Materie ist zunächst die Bewegung gegeben, d. h. die Geschwindigkeit $\mathfrak{q}$, welcher der Vierervektor $(\mathfrak{G}_\mu)$

$$(\mathfrak{G}_\mu) = \left(\frac{dx_\mu}{\sqrt{\sum dx_\sigma^2}}\right) \quad \ldots\ldots (49)$$

mit den Komponenten

$$\left.\begin{aligned} \mathfrak{G}_1 &= \frac{\mathfrak{q}_x}{\sqrt{c^2-\mathfrak{q}^2}} \\ &\text{- - - - - - -} \\ \mathfrak{G}_4 &= \frac{ic}{\sqrt{c^2-\mathfrak{q}^2}} \end{aligned}\right\} (49a)$$

entspricht. Ferner existiert im Raume ein durch die Feldvektoren $\mathfrak{f}$ und $\mathfrak{n}$, ~~charakterisiert~~ vierdimensional durch die Sechservektoren $(\mathfrak{F}_{\mu\nu})$ und $(\mathfrak{F}^*_{\mu\nu})$ charakterisiertes Vakuum-Feld. Aus diesen und dem Vierervektor der Geschwindigkeit lassen sich zwei Vierervektoren $(\mathfrak{F}_\mu)$ und $(\mathfrak{F}^*_\mu)$ ~~bilden~~, welche wir die Vierervektoren der elektromotorischen bezw. magnetomotorischen Kraft $(\mathfrak{F}_\mu^{(e)})$ bezw. $(\mathfrak{F}_\mu^{(m)})$ ~~nennen~~; sie seien definiert durch die Gleichungen

$$\left.\begin{aligned} (\mathfrak{F}_\mu^{(e)}) &= (\mathfrak{F}_{\mu\nu})(\mathfrak{G}_\nu) \\ \mathfrak{F}_1^{(e)} &= \frac{1}{\sqrt{1-\frac{\mathfrak{q}^2}{c^2}}}\left(\mathfrak{n}_x + \frac{\mathfrak{q}_y}{c}\mathfrak{f}_z - \frac{\mathfrak{q}_z}{c}\mathfrak{f}_y\right) \\ &\text{- - - - - - - - - - - - -} \\ \mathfrak{F}_4^{(e)} &= \frac{i}{\sqrt{1-\frac{\mathfrak{q}^2}{c^2}}}\left(\frac{\mathfrak{q}_x}{c}\mathfrak{n}_x + \frac{\mathfrak{q}_y}{c}\mathfrak{n}_y + \frac{\mathfrak{q}_z}{c}\mathfrak{n}_z\right) \end{aligned}\right\} (50)$$

$$\left.\begin{aligned} (\mathfrak{F}_\mu^{(m)}) &= -i(\mathfrak{F}^*_{\mu\nu})(\mathfrak{G}_\nu) \\ \mathfrak{F}_1^{(m)} &= \frac{1}{\sqrt{1-\frac{\mathfrak{q}^2}{c^2}}}\left(\mathfrak{f}_x - \frac{\mathfrak{q}_y}{c}\mathfrak{n}_z + \frac{\mathfrak{q}_z}{c}\mathfrak{n}_y\right) \\ &\text{- - - - - - - - - - - - -} \\ \mathfrak{F}_4^{(m)} &= \frac{i}{\sqrt{1-\frac{\mathfrak{q}^2}{c^2}}}\left(\frac{\mathfrak{q}_x}{c}\mathfrak{f}_x + \frac{\mathfrak{q}_y}{c}\mathfrak{f}_y + \frac{\mathfrak{q}_z}{c}\mathfrak{f}_z\right) \end{aligned}\right\} (51)$$

Wir haben jetzt noch den elektromagnetischen Zustand der ponderablen Materie in solcher Weise festzulegen, dass nur vierdimensional kovariante ~~Bestimmun~~ Gebilde herangezogen werden. Dabei gehen wir so vor, dass wir wieder diejenigen Dinge säuberlich auseinanderhalten, welche vom physikalischen Standpunkte aus betrachtet eine ~~unabhängige~~ selbständige Existenz besitzen.

a) <u>Elektrische Polarisation.</u>

Die elektrische Polarisation ist in ruhenden Medien durch den Vektor $\mathfrak{p}$ dargestellt. Diesem lassen wir einen Vierervektor $(\mathfrak{P}_\mu)$ entsprechen, mit der zusätzlichen Bestimmung, dass im Falle der Ruhe (mitbewegtes System)

$$\mathfrak{P}_1 = \mathfrak{p}_x,\ \mathfrak{P}_2 = \mathfrak{p}_y,\ \mathfrak{P}_3 = \mathfrak{p}_z,\ \mathfrak{P}_4 = 0$$

sein soll. Dieser Vierervektor erfüllt für das mitbewegte System, also überhaupt die Bedingung

$$(\mathfrak{P}_\mu)(\mathfrak{G}_\mu) = 0. \quad \ldots (52)$$

Mit Rücksicht auf diese Gleichung können wir, indem wir die Bedeutung der Zeichen ~~$\mathfrak{P}$~~ $\mathfrak{p}_x$ etc. ausdehnen, für beliebig bewegte Körper setzen

$$\left.\begin{aligned}\mathfrak{P}_1 &= \mathfrak{p}_x\\ \mathfrak{P}_2 &= \mathfrak{p}_y\\ \mathfrak{P}_3 &= \mathfrak{p}_z\\ \mathfrak{P}_4 &= i\,(\mathfrak{p}\frac{\mathfrak{q}}{c})\end{aligned}\right\} \qquad (53)$$

Further, by forming the outer product of the four-vector $\mathfrak{P}_\mu$ with the four-vector of the velocity G_ν, we can form the six-vector

$$(\mathfrak{P}_{\mu\nu}) = (G_\mu)\,(\mathfrak{P}_\nu - (G_\nu)\,(\mathfrak{P}_\mu) \qquad \ldots (54)$$

of electric polarization, with the components

$$\left.\begin{aligned}\mathfrak{P}_{23} &= \frac{1}{\sqrt{1-\frac{q^2}{c^2}}}\left(\frac{\mathfrak{q}_y}{c}\mathfrak{p}_z - \frac{\mathfrak{q}_z}{c}\mathfrak{p}_y\right) \quad \text{etc.}\\ \mathfrak{P}_{14} &= \frac{i}{\sqrt{1-\frac{q^2}{c^2}}}\left\{\frac{\mathfrak{q}_x}{c}\left(\frac{\mathfrak{q}}{c},\mathfrak{p}\right) - \mathfrak{p}_x\right\} \quad \text{etc}\end{aligned}\right\} \qquad (55)$$

b. Magnetic Polarization

In the same manner we define a four-vector $(\mathfrak{M}_\mu)$ of magnetic polarization that satisfies the condition

$$(\mathfrak{M}_\mu)\,(G_\mu) = 0 \qquad \ldots (56)$$

and the components of which can be represented by the three-dimensional $\mathfrak{m}$ in the form

$$\left.\begin{aligned}\mathfrak{M}_1 &= \mathfrak{m}_x\\ &- - - -\\ &- - - -\\ \mathfrak{M}_4 &= i\,(\mathfrak{m},\frac{\mathfrak{q}}{c})\end{aligned}\right\} \qquad (57)$$

The corresponding six-vector $(\mathfrak{M}_{\mu\nu})$ of the magnetic polarization is defined by the equations

$$(\mathfrak{M}_{\mu\nu}) = (G_\mu)\,(\mathfrak{M}_\nu) - (G_\nu)\,(\mathfrak{M}_\mu) \qquad \ldots (58)$$

$$\left.\begin{aligned}\mathfrak{M}_{23} &= \frac{1}{\sqrt{1-\frac{q^2}{c^2}}}\left(\frac{\mathfrak{q}_y}{c}\mathfrak{m}_z - \frac{\mathfrak{q}_z}{c}\mathfrak{m}_y\right) \quad \text{etc.}\\ \mathfrak{M}_{14} &= \frac{i}{\sqrt{1-\frac{q^2}{c^2}}}\left\{\frac{\mathfrak{q}_x}{c}\left(\frac{\mathfrak{q}}{c},\mathfrak{m}\right) - \mathfrak{m}_x\right\} \quad \text{etc}\end{aligned}\right\} \qquad (59)$$

c. Conduction Current

With respect to a comoving system, the density of a conduction current is defined by three components $\mathfrak{i}_x$, $\mathfrak{i}_y$, $\mathfrak{i}_z$. We represent it by a four-vector $(\mathfrak{J}_\mu^{(l)})$ whose fourth component vanishes for a comoving system. Thus, we have

$$(\mathfrak{J}_\mu^{(l)})(G_\mu) = 0 \qquad \ldots (60)$$

and, extending again the meaning of $\mathfrak{i}$,

$$\left.\begin{aligned}\mathfrak{J}_1 &= \mathfrak{i}_x\\ &- - - - -\\ &- - - - -\\ \mathfrak{J}_4 &= i\,(\mathfrak{i},\frac{\mathfrak{q}}{c})\end{aligned}\right\} \qquad (61)$$

d. Convection Current

To this current there corresponds a four-vector $(\mathfrak{J}_\mu^k)$ whose components are proportional to those of the four-vector of velocity. We set

$$(\mathfrak{J}_\mu^k) = \rho_0(G_\mu), \qquad \ldots (62)$$

where ρ_0 denotes a scalar, the "rest-density." Expressed in full, we have

$$\left.\begin{aligned} \mathfrak{P}_1 &= \mathfrak{p}_x \\ \mathfrak{P}_2 &= \mathfrak{p}_y \\ \mathfrak{P}_3 &= \mathfrak{p}_z \\ \mathfrak{P}_4 &= i\left(\mathfrak{p}\,\frac{\mathfrak{q}}{c}\right) \end{aligned}\right\} \quad (53)$$

Durch äussere Multiplikation des Vierervektors $(\mathfrak{P}_\mu)$ mit dem Vierervektor der Geschwindigkeit (G_ν) können wir ferner den Sechservektor

$$(\mathfrak{P}_{\mu\nu}) = (G_\mu)(\mathfrak{P}_\nu) - (G_\nu)(\mathfrak{P}_\mu) \quad \ldots\ldots (54)$$

der elektrischen Polarisation bilden mit den Komponenten

$$\left.\begin{aligned} \mathfrak{P}_{23} &= \frac{1}{\sqrt{1-\frac{q^2}{c^2}}}\left(\frac{\mathfrak{q}_y}{c}\mathfrak{p}_z - \frac{\mathfrak{q}_z}{c}\mathfrak{p}_y\right) \text{ etc.} \\ \mathfrak{P}_{14} &= \frac{i}{\sqrt{1-\frac{q^2}{c^2}}}\left\{\frac{\mathfrak{q}_x}{c}\left(\frac{\mathfrak{q}}{c}, \mathfrak{p}\right) - \mathfrak{p}_x\right\} \text{ etc} \end{aligned}\right\} \quad (55)$$

b) Magnetische Polarisation

In gleicher Weise definieren wir einen Vierervektor $(\mathfrak{M}_\mu)$ der magnetischen Polarisation, der die Bedingung

$$(\mathfrak{M}_\mu)(G_\mu) = 0 \quad \ldots\ldots (56)$$

erfüllt, und dessen Komponenten sich durch das dreidimensionale $\mathfrak{m}$ in der Form

$$\left.\begin{aligned} \mathfrak{M}_1 &= \mathfrak{m}_x \\ &\text{-- -- -- --} \\ &\text{-- -- -- --} \\ \mathfrak{M}_4 &= i\left(\mathfrak{m}, \frac{\mathfrak{q}}{c}\right) \end{aligned}\right\} \quad (57)$$

darstellen lassen. Der entsprechende Sechservektor $(\mathfrak{M}_{\mu\nu})$ der magnetischen Polarisation wird durch die Gleichungen

$$(\mathfrak{M}_{\mu\nu}) = (G_\mu)(\mathfrak{M}_\nu) - (G_\nu)(\mathfrak{M}_\mu) \quad \ldots (58)$$

$$\left.\begin{aligned} \mathfrak{M}_{23} &= \frac{1}{\sqrt{1-\frac{q^2}{c^2}}}\left(\frac{\mathfrak{q}_y}{c}\mathfrak{m}_z - \frac{\mathfrak{q}_z}{c}\mathfrak{m}_y\right) \text{ etc.} \\ \mathfrak{M}_{14} &= \frac{i}{\sqrt{1-\frac{q^2}{c^2}}}\left\{\frac{\mathfrak{q}_x}{c}\left(\frac{\mathfrak{q}}{c}, \mathfrak{m}\right) - \mathfrak{m}_x\right\} \text{ etc} \end{aligned}\right\} \quad (59)$$

definiert.

c) Leitungsstrom

Die Dichte des Leitungsstromes ist bezüglich eines mitbewegten Systems durch drei Komponenten i_x, i_y, i_z definiert. Wir ~~ersetzen~~ stellen ihn durch einen Vierervektor $(\mathfrak{J}_\mu^{(l)})$ dar, dessen vierte Komponente für ein mitbewegtes System verschwindet. Man hat also

$$(\mathfrak{J}_\mu^{l})(G_\mu) = 0 \quad \ldots\ldots (60)$$

und, indem wir wieder die Bedeutung von i erweitern

$$\left.\begin{aligned} \mathfrak{J}_1 &= i_x \\ &\text{-- -- -- --} \\ &\text{-- -- -- --} \\ \mathfrak{J}_4 &= i\left(i, \frac{\mathfrak{q}}{c}\right) \end{aligned}\right\} \quad (61)$$

d) Konvektionsstrom

Diesem entspricht ein Vierervektor $(\mathfrak{J}_\mu^{k})$, dessen Komponenten denen des Vierervektors der Geschwindigkeit proportional sind. Wir setzen

$$(\mathfrak{J}_\mu^{k}) = \varrho_0 (G_\mu), \quad \ldots (62)$$

wobei ϱ_0 einen Skalar, die „Ruhedichte" bedeutet. Ausführlich haben wir

$$\left.\begin{array}{l} \mathfrak{J}_i^k = \dfrac{\rho_0}{\sqrt{1-\dfrac{q^2}{c^2}}}\dfrac{\mathfrak{q}_x}{c} = \rho\dfrac{\mathfrak{q}_x}{c} \\ - - - - - - - - - - - - \\ \mathfrak{J}_4^k = \dfrac{i\rho_0}{\sqrt{1-\dfrac{q^2}{c^2}}} = i\rho \end{array}\right\} \qquad (63)$$

ρ denotes the density of the charge that generates the convection current, with respect to the coordinate system.—

Now we can easily specify equations in the four-dimensional, covariant form, which differ from equations (Ib) of §4 only in terms of second and higher order with respect to $\frac{\mathfrak{q}_x}{c}$, etc. We postulate

$$\left.\begin{array}{l} \left(\dfrac{\partial}{\partial x_\nu}\right)(\mathfrak{F}_{\mu\nu}+\mathfrak{P}_{\mu\nu}) = \dfrac{1}{c}(\mathfrak{J}_\mu^l) + \rho_0(G_\mu) \\ \left(\dfrac{\partial}{\partial x_\nu}\right)(\mathfrak{F}^*{}_{\mu\nu}+i\mathfrak{M}_{\mu\nu}) = 0 \end{array}\right\} \qquad (\mathrm{Ic})$$

Apart from the notation, these equations agree with Minkowski's field equations; they rest on no other assumption save the assumption that the medium may be treated as a continuum from an electromagnetic point of view.[87]

Now it remains for us to set up the equations that describe how the polarizations and currents in the matter depend on the field. We restrict ourselves to the case where the medium is isotropic at each of its space-time points for a comoving observer. Let equations of the following form hold at that point for such an observer

$$\mathfrak{p} = (\varepsilon-1)\,\mathfrak{e}$$
$$\mathfrak{m} = (\mu-1)\,\mathfrak{h}$$
$$\mathfrak{i}^l = \sigma\mathfrak{e}\,.$$

In the case where $\mathfrak{q} = 0$, these equations are equivalent to the four-dimensional vector equations

$$\left.\begin{array}{l} (\mathfrak{P}_\mu) = (\varepsilon-1)\,(\mathfrak{F}_\mu^e) \\ (\mathfrak{M}_\mu) = (\mu-1)\,(\mathfrak{F}_\mu^m) \\ (\mathfrak{J}_\mu^l) = \sigma\,(\mathfrak{F}_\mu^e) \end{array}\right\} ; \qquad (64)$$

hence they express the same thing as these equations do for an arbitrary justified system of reference. We need only one more equation, which expresses $(\mathfrak{J}_\mu^k)$ in terms of the field quantities. If we multiply the first of equations (Ic) by (G_μ)we obtain, in light of (60), (62), and 49,

$$(G_\mu)\left(\frac{\partial(\mathfrak{F}_{\mu\nu}+\mathfrak{P}_{\mu\nu})}{\partial x_\nu}\right) = (\mathfrak{J}_\mu^k)\,(G_\mu) = \rho_0(G_\mu)\,(G_\mu) = -\rho_0$$

or

$$\rho_0 = -(G_\mu)\left(\frac{\partial(\mathfrak{F}_{\mu\nu}+\mathfrak{P}_{\mu\nu})}{\partial x_\nu}\right). \qquad \ldots (65)$$

In conjunction with (50), (64), and (54) and (62), this equation determines $(\mathfrak{J}_\mu^k)$ in terms of $(\mathfrak{F}_{\mu\nu})$; this equation is merely a consequence of the first of the vector equations (Ic) and therefore does not need to be added to the system of the fundamental equations.

$$\left.\begin{array}{l} \mathfrak{J}_1^K = \dfrac{\varrho_0}{\sqrt{1-\frac{q^2}{c^2}}}\,\dfrac{q_x}{c} = \varrho\,\dfrac{q_x}{c} \\ \text{- - - - - - - - - - - -} \\ \mathfrak{J}_4^K = \dfrac{i\varrho_0}{\sqrt{1-\frac{q^2}{c^2}}} = i\varrho \end{array}\right\} (63)$$

ϱ bedeutet die Dichte der den Konvektionsstrom verursachenden Ladung, auf das Koordinatensystem bezogen. —

Nun gelingt es leicht, Gleichungen in vierdimensional kovarianter Form anzugeben, welche sich von Gleichungen (Ib) des § 4 nur in Gliedern unterscheiden, die bezüglich $\frac{q_x}{c}$ etc. von zweiter und höherer Ordnung sind. Wir setzen an

$$\left.\begin{array}{l} \left(\dfrac{\partial}{\partial x_\nu}\right)(\mathfrak{F}_{\mu\nu} + \mathfrak{P}_{\mu\nu}) = \dfrac{1}{c}(\mathfrak{J}_\mu^l) + \varrho_0(\mathfrak{G}_\mu) \\ \left(\dfrac{\partial}{\partial x_\nu}\right)(\mathfrak{F}_{\mu\nu}^* + i\mathfrak{M}_{\mu\nu}) = 0 \end{array}\right\} (Ic)$$

Diese Gleichungen stimmen abgesehen von der Bezeichnungsweise mit Minkowskis Feldgleichungen überein. Sie ruhen auf keiner anderen Voraussetzung als auf der, dass das Medium in elektromagnetischer Beziehung als Kontinuum behandelt werden darf.

Nun müssen wir noch Gleichungen aufstellen, nach welchen in der Materie die Polarisationen und Ströme vom Felde abhängen. Wir beschränken uns dabei auf den Fall, dass das Medium in jedem seiner Raum-Zeit-Punkte für einen mitbewegten Beobachter isotrop ist. Für einen solchen mögen in jenem Punkte Gleichungen von der Form gelten

$$\mathfrak{p} = (\varepsilon - 1)\,\mathfrak{e}$$
$$\mathfrak{m} = (\mu - 1)\,\mathfrak{h}$$
$$\mathfrak{i}^l = \sigma\,\mathfrak{e}.$$

Mit diesen Gleichungen sind im Falle $q = 0$ die vierdimensionalen Vektorgleichungen

$$\left.\begin{array}{l} (\mathfrak{P}_\mu) = (\varepsilon-1)(\mathfrak{F}_\mu^e) \\ (\mathfrak{M}_\mu) = (\mu-1)(\mathfrak{F}_\mu^m) \\ (\mathfrak{J}_\mu^l) = \sigma(\mathfrak{F}_\mu^e) \end{array}\right\} (64)$$

gleichbedeutend; sie drücken daher dasselbe wie jene Gleichungen für ein beliebiges berechtigtes Bezugssystem aus. Es fehlt nur noch eine Gleichung, welche $(\mathfrak{J}_\mu^K)$ durch die Feldgrössen ausdrückt. Multiplizieren wir die erste der Gleichungen (Ic) mit $(\mathfrak{G}_\mu)$, so erhalten wir mit Rücksicht auf (60), (62) und 49

$$\sum_\nu (\mathfrak{G}_\mu)\left(\frac{\partial(\mathfrak{F}_{\mu\nu} + \mathfrak{P}_{\mu\nu})}{\partial x_\nu}\right) = (\mathfrak{J}_\mu^K)(\mathfrak{G}_\mu) = \varrho_0(\mathfrak{G}_\mu)(\mathfrak{G}_\mu) = -\varrho_0$$

oder

$$\varrho_0 = -(\mathfrak{G}_\mu)\left(\frac{\partial(\mathfrak{F}_{\mu\nu} + \mathfrak{P}_{\mu\nu})}{\partial x_\nu}\right) \dots (65)$$

Diese Gleichung bestimmt in Verbindung mit (50), (64), (54) und (62) $(\mathfrak{J}_\mu^K)$ durch $(\mathfrak{F}_{\mu\nu})$; sie ist eine blosse Konsequenz der ersten der Vektorgleichungen (Ic) und braucht daher dem System der Fundamentalgleichungen nicht angereiht zu werden.

Considering that by virtue of (64) and the definitions given at the beginning of the §, the vectors $(\mathfrak{P}_{\mu\nu})$, $(\mathfrak{M}_{\mu\nu})$, and $(\mathfrak{S}^{l}_{\mu})$ can be expressed in terms of $(\mathfrak{F}_{\mu\nu})^*$ and (G_μ), equations (Ic) contain all of the conditions that the electromagnetic field has to satisfy if $\mathfrak{q}$, ε, μ, σ are given as functions of *x, y, z, t*. For the system (Ic) contains 8 condition equations for ρ_0 and the 6 components $\mathfrak{F}_{\mu\nu}$. The surplus condition expresses the absence of magnetic charges.

It should be noted that our equations differ only formally from those of Minkowski.[88] We prefer this system because the quantities appearing in it have a <much> more direct physical meaning, and because this system lends itself better to an extension to cases in which the connections between field strength and polarization are less simple. Finally, it also turns out that the derivation of the conservation laws (and the ponderomotive forces) becomes simpler.

We now write the equations of the theory in the form of the three-dimensional vector calculus by splitting both of the six-vectors of the polarizations $(\mathfrak{P}_{\mu\nu})$ and $(\mathfrak{M}_{\mu\nu})$ into two ordinary vectors $\overset{*}{\mathfrak{p}}$, $\overset{**}{\mathfrak{p}}$ and $\overset{*}{\mathfrak{m}}$, $\overset{**}{\mathfrak{m}}$, respectively.* Instead of the field equations (Ic), we then obtain

$$\left.\begin{aligned}
&\operatorname{rot}(\mathfrak{h}+\overset{*}{\mathfrak{p}}) - \frac{1}{c}\frac{\partial(\mathfrak{e}+\overset{**}{\mathfrak{p}})}{\partial t} = \frac{1}{c}\mathfrak{i} + \rho\frac{\mathfrak{q}}{c} \qquad && \operatorname{rot}(\mathfrak{e}-\overset{*}{\mathfrak{m}}) + \frac{1}{c}\frac{\partial(\mathfrak{h}+\overset{**}{\mathfrak{m}})}{\partial t} = 0\\
&\operatorname{div}(\mathfrak{e}+\overset{**}{\mathfrak{p}}) = \frac{1}{c}\left(\mathfrak{i}, \frac{\mathfrak{q}}{c}\right) + \rho && \operatorname{div}(\mathfrak{h}+\overset{**}{\mathfrak{m}}) = 0
\end{aligned}\right\} \quad \text{(Id)}$$

The vectors of polarization are connected with those of the electromagnetic field by the following chain of relations

$$\left.\begin{aligned}
&\overset{*}{\mathfrak{e}} = \frac{\mathfrak{e} + \left[\frac{\mathfrak{q}}{c}, \mathfrak{h}\right]}{\sqrt{1-\frac{q^2}{c^2}}} \qquad && \overset{*}{\mathfrak{h}} = \frac{\mathfrak{h} - \left[\frac{\mathfrak{q}}{c}, \mathfrak{e}\right]}{\sqrt{1-\frac{q^2}{c^2}}}\\
&\mathfrak{p} = (\varepsilon - 1)\,\overset{*}{\mathfrak{e}} && \mathfrak{m} = (\mu - 1)\,\overset{*}{\mathfrak{h}}\\
&\overset{*}{\mathfrak{p}} = \frac{\left[\frac{\mathfrak{q}}{c}, \mathfrak{p}\right]}{\sqrt{1-\frac{q^2}{c^2}}} && \overset{*}{\mathfrak{m}} = \frac{\left[\frac{\mathfrak{q}}{c}, \mathfrak{m}\right]}{\sqrt{1-\frac{q^2}{c^2}}}\\
&\overset{**}{\mathfrak{p}} = \frac{\mathfrak{p} - \frac{\mathfrak{q}}{c}\left(\frac{\mathfrak{q}}{c}, \mathfrak{p}\right)}{\sqrt{1-\frac{q^2}{c^2}}} && \overset{**}{\mathfrak{m}} = \frac{\mathfrak{m} - \frac{\mathfrak{q}}{c}\left(\frac{\mathfrak{q}}{c}, \mathfrak{m}\right)}{\sqrt{1-\frac{q^2}{c^2}}}
\end{aligned}\right\} \quad (66)$$

To this is to be added for the conduction current the relation

$$\mathfrak{i} = \sigma\mathfrak{e}^* \qquad \ldots (66\text{a})$$

Boundary conditions

The second and fourth of equations (Id) lead to the equations for the boundary between two media

---

* i.e., we set $\overset{*}{\mathfrak{p}}_x = \mathfrak{P}_{23}$, $-i\overset{**}{\mathfrak{p}}_x = \mathfrak{P}_{14}$ etc and $\overset{*}{\mathfrak{m}}_x = \mathfrak{M}_{23}$, $-i\overset{**}{\mathfrak{m}}_x = \mathfrak{M}_{14}$.

Mit Rücksicht darauf, dass sich vermöge (64) und der am Anfang des § gegebenen Definitionen die Vektoren $(\mathfrak{P}_{\mu\nu})$, $(\mathfrak{M}_{\mu\nu})$ und $(\mathfrak{J}_\mu^l)$ durch $(\mathfrak{F}_{\mu\nu})$ und $(\mathfrak{G}_\mu)$ ausdrücken lassen, enthalten die Gleichungen (I_c) alle Bedingungen, denen das elektromagnetische Feld zu genügen hat, wenn $\eta, \varepsilon, \mu, \sigma$ als Funktion von x, y, z, t gegeben sind. Denn das System (I_c) enthält 8 Bedingungsgleichungen für ϱ_0 und die 6 Komponenten $\mathfrak{F}_{\mu\nu}$. Die überzählige Bedingung drückt das Fehlen von magnetischen Ladungen aus. ~~Es ist zu~~

Es ist zu bemerken, dass unsere Gleichungen sich von denjenigen Minkowskis nur formal unterscheiden. Wir geben diesem System den Vorzug, weil ~~da~~ die in ihm auftretenden Grössen eine ~~viel~~ unmittelbarere physikalische Bedeutung haben, ~~In dreidimensionaler Schreib- und sich für eine~~ und weil sich dies System besser eignet für eine Erweiterung auf ~~den~~ Fälle, in denen die Zusammenhänge zwischen Feldstärke und Polarisationen minder einfach sind. Endlich zeigt sich auch, dass die Ableitung der Erhaltungssätze (bezw. der ponderomotorischen Kräfte) eine Vereinfachung erfährt.

Wir schreiben nun die Gleichungen der Theorie in der Form dreidimensionaler Vektorenrechnung, indem wir die Sechservektoren der Polarisationen $(\mathfrak{P}_{\mu\nu})$ und $(\mathfrak{M}_{\mu\nu})$ je in zwei gewöhnliche Vektoren $\mathfrak{p}^*, \mathfrak{p}^{**}$ und $\mathfrak{m}^*, \mathfrak{m}^{**}$ spalten.[x]

Anstelle der Feldgleichungen (I_c) erhalten wir dann

$$\left.\begin{aligned} \mathrm{rot}(\mathfrak{h}+\mathfrak{p}^*) - \frac{1}{c}\frac{\partial(\mathfrak{e}+\mathfrak{p}^{**})}{\partial t} &= \frac{1}{c}\mathfrak{i} + \varrho\frac{\mathfrak{q}}{c} \qquad & \mathrm{rot}(\mathfrak{e}-\mathfrak{m}^*) + \frac{1}{c}\frac{\partial(\mathfrak{h}+\mathfrak{m}^{**})}{\partial t} &= 0 \\ \mathrm{div}(\mathfrak{e}+\mathfrak{p}^{**}) &= \frac{1}{c}\left(\mathfrak{i}, \frac{\mathfrak{q}}{c}\right) + \varrho & \mathrm{div}(\mathfrak{h}+\mathfrak{m}^{**}) &= 0 \end{aligned}\right\} (I_d)$$

Die Vektoren der Polarisation ~~und des Leitungsstromes~~ sind mit denen des elektromagnetischen Feldes durch folgende Kette von Relationen verbunden:

$$\left.\begin{aligned} \mathfrak{e}^* &= \frac{\mathfrak{e} + \left[\frac{\mathfrak{q}}{c}, \mathfrak{h}\right]}{\sqrt{1-\frac{q^2}{c^2}}} \qquad & \mathfrak{h}^* &= \frac{\mathfrak{h} - \left[\frac{\mathfrak{q}}{c}, \mathfrak{e}\right]}{\sqrt{1-\frac{q^2}{c^2}}} \\ \mathfrak{p} &= (\varepsilon-1)\,\mathfrak{e}^* & \mathfrak{m} &= (\mu-1)\,\mathfrak{h}^* \\ \mathfrak{p}^* &= \frac{\left[\frac{\mathfrak{q}}{c}, \mathfrak{p}\right]}{\sqrt{1-\frac{q^2}{c^2}}} & \mathfrak{m}^* &= \frac{\left[\frac{\mathfrak{q}}{c}, \mathfrak{m}\right]}{\sqrt{1-\frac{q^2}{c^2}}} \\ \mathfrak{p}^{**} &= \frac{\mathfrak{p} - \frac{\mathfrak{q}}{c}\left(\frac{\mathfrak{q}}{c}, \mathfrak{p}\right)}{\sqrt{1-\frac{q^2}{c^2}}} & \mathfrak{m}^{**} &= \frac{\mathfrak{m} - \frac{\mathfrak{q}}{c}\left(\frac{\mathfrak{q}}{c}, \mathfrak{m}\right)}{\sqrt{1-\frac{q^2}{c^2}}} \end{aligned}\right\} (66)$$

Dazu kommt für den Leitungsstrom die Relation

$$\mathfrak{i} = \sigma\,\mathfrak{e}^* \quad \ldots (66a)$$

Grenzbedingungen.

Aus der zweiten und vierten der Gleichungen (I_d) folgen für die Grenze zwischen zwei Medien die Gleichungen

$$(\mathfrak{e}+\mathfrak{p}^{**})_{n_1} + (\mathfrak{e}+\mathfrak{p}^{**})_{n_2} = \delta$$

$$(\mathfrak{h}+\mathfrak{m}^{**})_{n_1} + (\mathfrak{h}+\mathfrak{m}^{**})_{n_2} = 0,$$

wobei n_1 bezw. n_2 die Richtung der nach dem Innern des betreffenden Mediums gerichtete Normale, δ das Volumintegral der Skalars $\frac{1}{c}\left(\mathfrak{i}, \frac{\mathfrak{q}}{c}\right) + \varrho$ über ~~die Fläche~~ Grenzschicht pro Flächeneinheit derselben bedeutet.

[x] d. h. wir setzen

$\mathfrak{p}^*_x = \mathfrak{P}_{23}$ etc und $\mathfrak{m}^*_x = \mathfrak{M}_{23}$

$-i\mathfrak{p}^{**}_x = \mathfrak{P}_{14}$ $\quad$ $-i\mathfrak{m}^{**}_x = \mathfrak{M}_{14}$

$$\left.\begin{aligned}(\mathfrak{e}+\overset{**}{\mathfrak{p}})_{n_2}-(\mathfrak{e}+\overset{**}{\mathfrak{p}})_{n_1}&=\delta\\ (\mathfrak{h}+\overset{**}{\mathfrak{m}})_{n_2}-(\mathfrak{h}+\overset{**}{\mathfrak{m}})_{n_1}&=0\end{aligned}\right\} \tag{67}$$

where n denotes the direction of the normal drawn from the first to the second body of the boundary surface and δ denotes the volume integral of the density ρ of the convection current per unit surface area of the boundary layer.

We arrive at the rest of the boundary conditions, for the case where $\mathfrak{q}$ remains constant in the transitional layer, by a route that should be indicated in the first of equations (Id). In the boundary layer the expressions

$$\operatorname{rot}(\mathfrak{h}+\overset{*}{\mathfrak{p}})-\frac{1}{c}\frac{\partial(\mathfrak{e}+\overset{**}{\mathfrak{p}})}{\partial t}-\frac{\mathfrak{q}}{c}\rho \qquad \text{(according to the first equation (Ib)}$$

and

$$\frac{\partial(\mathfrak{e}+\overset{**}{\mathfrak{p}})}{\partial t}+(\mathfrak{q}\Delta)(\mathfrak{e}+\overset{**}{\mathfrak{p}}) \qquad \text{(temporal change for a comoving particle)}$$

remain finite, and thus also the expression

$$\operatorname{rot}(\mathfrak{h}+\overset{*}{\mathfrak{p}})+\left(\frac{\mathfrak{q}}{c}\Delta\right)(\mathfrak{e}+\overset{**}{\mathfrak{p}})-\frac{\mathfrak{q}}{c}\rho.$$

From this we can conclude with the help of the second of equations (Id) that in the boundary layer the tangential components of the vector

$$\mathfrak{h}+\overset{*}{\mathfrak{p}}-\left[\frac{\mathfrak{q}}{c},\mathfrak{e}+\overset{**}{\mathfrak{p}}\right]$$

remain constant. The same can be deduced for the vector

$$\mathfrak{e}-\overset{*}{\mathfrak{m}}+\left[\frac{\mathfrak{q}}{c},\mathfrak{h}+\overset{**}{\mathfrak{m}}\right]$$

from the third of equations (Id). Thus, if one denotes a direction parallel to the boundary surface by l, one obtains the boundary conditions

$$\left.\begin{aligned}\left\{\mathfrak{h}+\overset{*}{\mathfrak{p}}-\left[\frac{\mathfrak{q}}{c},\mathfrak{e}+\overset{**}{\mathfrak{p}}\right]\right\}_{l_1}&=\left\{\mathfrak{h}+\overset{*}{\mathfrak{p}}-\left[\frac{\mathfrak{q}}{c},\mathfrak{e}+\overset{**}{\mathfrak{p}}\right]\right\}_{l_2}\\ \left\{\mathfrak{e}-\overset{*}{\mathfrak{m}}+\left[\frac{\mathfrak{q}}{c},\mathfrak{h}+\overset{**}{\mathfrak{m}}\right]\right\}_{l_1}&=\left\{\mathfrak{e}-\overset{*}{\mathfrak{m}}+\left[\frac{\mathfrak{q}}{c},\mathfrak{h}+\overset{**}{\mathfrak{m}}\right]\right\}_{l_2}\end{aligned}\right\} \tag{68}$$

§20. Momentum, Energy, and Ponderomotive Forces

The General Form of the Momentum-Energy Law

We saw in §18 that in the electrodynamics of the vacuum the energy balance (48) is based on the existence of a symmetric tensor $T_{\mu\nu}$, the components of which are given by the following schema in accordance with their physical meanings:

$$\left.\begin{matrix} p_{xx} & p_{xy} & p_{xz} & ic\mathfrak{g}_x \\ p_{yx} & p_{yy} & p_{yz} & ic\mathfrak{g}_y \\ p_{zx} & p_{zy} & p_{zz} & ic\mathfrak{g}_z \\ \frac{i}{c}\mathfrak{s}_x & \frac{i}{c}\mathfrak{s}_y & \frac{i}{c}\mathfrak{s}_z & -\eta \end{matrix}\right\} \tag{69}$$

p_{xx} etc. denote the Maxwell stresses, $\mathfrak{g}$ the vector of the momentum density, $\mathfrak{s}$ the vector of the energy flow, and η the energy density. The circumstance that the tensor is symmetric leads to the relations

$$p_{xy}=p_{yx},\ \text{etc.} \qquad \ldots (70)$$

and

$$\mathfrak{g}=\frac{1}{c^2}\mathfrak{s}. \qquad \ldots (71)$$

$$\left.\begin{aligned}(\mathfrak{e}+\mathfrak{g}^{**})_{n2}-(\mathfrak{e}+\mathfrak{g}^{**})_{n1}&=\delta\\(\mathfrak{h}+\mathfrak{m}^{**})_{n2}-(\mathfrak{h}+\mathfrak{m}^{**})_{n1}&=0\end{aligned}\right\}\quad(67)$$

wobei n die Richtung der vom ersten nach dem zweiten Körper der Grenzfläche gezogene Normale, δ das Volumintegral der Dichte ϱ des Konvektionsstromes pro Flächeneinheit der Grenzschicht bedeutet.

Zu den übrigen Grenzbedingungen gelangen wir für den Fall, dass $\mathfrak{q}$ in der Übergangsschicht stetig bleibt, auf einem Wege, der an der ersten der Gleichungen (I d) angedeutet werden soll. Es bleiben in der Grenzschicht die Ausdrücke

$$\operatorname{rot}(\mathfrak{h}+\mathfrak{g}^{*})-\frac{1}{c}\frac{\partial(\mathfrak{e}+\mathfrak{g}^{**})}{\partial t}-\frac{\mathfrak{q}}{c}\varrho \quad \text{(nach der ersten Gleichung (I b))}$$

und

$$\frac{\partial\,\mathfrak{e}+\mathfrak{g}^{**}}{\partial t}+(\mathfrak{q}\Delta)(\mathfrak{e}+\mathfrak{g}^{**}) \quad \text{(zeitliche Änderung für ein mitbewegtes Teilchen)}$$

~~stetig~~ endlich, also auch der Ausdruck

$$\operatorname{rot}(\mathfrak{h}+\mathfrak{g}^{*})+\left(\frac{\mathfrak{q}}{c}\Delta\right)(\mathfrak{e}+\mathfrak{g}^{**})-\frac{\mathfrak{q}}{c}\varrho.$$

Hieraus, ~~sowie daraus, dass nach der zweiten~~ kann unter Benutzung der zweiten der Gleichungen (I d) gefolgert werden, dass in der Grenzschicht die Tangentialkomponenten des Vektors

$$\mathfrak{h}+\mathfrak{g}^{*}-\left[\frac{\mathfrak{q}}{c},\mathfrak{e}+\mathfrak{g}^{**}\right]$$

stetig bleiben. Gleiches ist aus der dritten der Gleichungen (I d) für den Vektor

$$\mathfrak{e}-\mathfrak{m}^{*}+\left[\frac{\mathfrak{q}}{c},\mathfrak{h}+\mathfrak{m}^{**}\right]$$

abzuleiten. ~~sodass man~~ Bezeichnet man also mit l eine der Grenzfläche parallele Richtung, so erhält man die Grenzbedingungen

$$\left.\begin{aligned}\left\{\mathfrak{h}+\mathfrak{g}^{*}-\left[\frac{\mathfrak{q}}{c},\mathfrak{e}+\mathfrak{g}^{**}\right]\right\}_{l_1}&=\left\{\mathfrak{h}+\mathfrak{g}^{*}-\left[\frac{\mathfrak{q}}{c},\mathfrak{e}+\mathfrak{g}^{**}\right]\right\}_{l_2}\\\left\{\mathfrak{e}-\mathfrak{m}^{*}+\left[\frac{\mathfrak{q}}{c},\mathfrak{h}+\mathfrak{m}^{**}\right]\right\}_{l_1}&=\left\{\mathfrak{e}-\mathfrak{m}^{*}+\left[\frac{\mathfrak{q}}{c},\mathfrak{h}+\mathfrak{m}^{**}\right]\right\}_{l_2}\end{aligned}\right\}\quad(68).$$

§20.

Impuls, Energie und ponderomotorische Kräfte.

~~Allgemeine Form des Impuls-Energie-Satzes~~

In §18 sahen ~~wir~~, dass in der Elektrodynamik des Vakuums die Energiebilanz ~~durch~~ (48) auf der Existenz ~~eines~~ symmetrischen Tensors $T_{\mu\nu}$ beruht, dessen Komponenten ihrer physikalischen Bedeutung nach durch folgendes Schema gegeben sind:

$$\left.\begin{matrix}p_{xx} & p_{xy} & p_{xz} & ic\mathfrak{g}_x\\ p_{yx} & p_{yy} & p_{yz} & ic\mathfrak{g}_y\\ p_{zx} & p_{zy} & p_{zz} & ic\mathfrak{g}_z\\ \frac{i}{c}\mathfrak{f}_x & \frac{i}{c}\mathfrak{f}_y & \frac{i}{c}\mathfrak{f}_z & -\eta,\end{matrix}\right\}\quad(69)$$

p_{xx} etc. bedeuten dabei die Maxwell'schen Spannungen, $\mathfrak{g}$ den Vektor der Impulsdichte, $\mathfrak{f}$ den Vektor der Energieströmung, η die Energiedichte. Der Umstand, dass der Tensor symmetrisch ist, bringt die Beziehungen

$$p_{xy}=p_{yx} \text{ etc.} \quad \ldots \quad ~~69~~\ (70)$$

und

$$\mathfrak{g}=\frac{1}{c^2}\mathfrak{f} \quad \ldots\ldots \quad (71)$$

~~Der Umstand, dass dieser~~ Gleichung (71) hängt aufs Engste damit zusammen,

Equation (71) is closely related to the circumstance that, according to the theory of relativity, an inertial mass must be ascribed to energy. For this is connected with the fact that the energy flow is always associated with a momentum.

If we form the divergence of $(T_{\mu\nu})$, we obtain a four-vector (K_μ)

$$(K_\mu) = -\left(\frac{\partial}{\partial x_\nu}\right)(T_{\mu\nu}), \qquad \ldots (48a)$$

whose spatial components K_1, K_2, K_3 are the components of the force that the field exerts, per unit volume, on the carriers of the electric masses, while K_4 signifies the energy multiplied by $\frac{i}{c}$ that the field delivers to the carriers of the electric masses per unit volume and unit time.

The general validity of the conservation laws and of the law of the inertia of energy, which has already been derived in §14, suggest that the relations (69), (70), (71), and (48a) are to be ascribed a *general* significance, even though they were obtained in a very special case. We owe this generalization, which is the most important new advance in the theory of relativity, to the investigations of Minkowski, Abraham, Planck, and Laue. [89] To every kind of material process we might study, we have to assign a symmetric tensor $(T_{\mu\nu})$, the components of which have the physical meaning indicated in (69). Then equation (48a) must always be satisfied. Each time, the problem to be solved consists in finding out how $(T_{\mu\nu})$ is to be formed from the variables characterizing the processes studied. If several processes that can be separated from one another as regards energy occur in the same space, then we have to assign to each such individual process its own stress-energy tensor $(T^{(1)}_{\mu\nu})$, etc., and to set $(T_{\mu\nu})$ equal to the sum of these individual tensors.

The Stress-Energy Tensor of Electromagnetic Processes

The electromagnetic states investigated in the last § are composed of three kinds of states that, according to our current knowledge, are energetically completely separable, namely, of

1) the electromagnetic states of the vacuum
2) the states of electrical polarization of the matter
3) the states of magnetic polarization of the matter;

we assign to them, respectively, the stress-energy tensors $(T^0_{\mu\nu})$, $(T^e_{\mu\nu})$, and $(T^m_{\mu\nu})$.

$(T^0_{\mu\nu})$ has the same expression as in the electrodynamics of the vacuum. Hence we have according to (48b)

$$(T^0_{\mu\nu}) = \frac{1}{2}\{(\mathfrak{F}_{\mu\sigma})(\mathfrak{F}_{\nu\sigma}) - (\mathfrak{F}^*{}_{\mu\sigma})(\mathfrak{F}^*{}_{\nu\sigma})\} \qquad \ldots (72)$$

The tensor $T^{(e)}_{\mu\nu}$ has been determined earlier for the case of rest ($\mathfrak{q} = 0$). For that case, the p^e_{xx} to p^e_{zz} are obtained from (10a). Furthermore, for the case of rest we must have $s^e = \mathfrak{g}^e = 0$, $w^e = \frac{1}{\varepsilon - 1}\frac{\mathfrak{g}^2}{2}$. This yields $(T^e_{\mu\nu})$ uniquely for the case of motion as

$$(T^e_{\mu\nu}) = \frac{1}{\varepsilon - 1}\{(\mathfrak{P}_{\mu\sigma})(\mathfrak{P}_{\nu\sigma}) - \frac{1}{4}(\delta_{\mu\nu})(\mathfrak{P}_{\sigma\tau})(\mathfrak{P}_{\sigma\tau})\}. \qquad \ldots (73)$$

Analogously, we must have

$$(T^m_{\mu\nu}) = \frac{1}{\mu - 1}\{(\mathfrak{M}_{\mu\sigma})(\mathfrak{M}_{\nu\sigma}) - \frac{1}{4}(\delta_{\mu\nu})(\mathfrak{M}_{\sigma\tau})(\mathfrak{M}_{\sigma\tau})\} \qquad \ldots (74)$$

dass nach der Relativitätstheorie der Energie eine träge Masse zuzuschreiben ist. Denn damit hängt es zusammen, dass mit dem Energiestrom stets ein Impuls verknüpft ist.

Bilden wir die Divergenz von $(T_{\mu\nu})$, so erhalten wir einen Vierervektor (K_μ)

$$(K_\mu) = -\left(\frac{\partial}{\partial x_\nu}\right)(T_{\mu\nu}), \quad \ldots \ldots (48a)$$

dessen räumliche Komponenten K_1, K_2, K_3 die Komponenten der vom Felde auf die Träger der elektrischen Massen pro Volumeneinheit ausgeübten Kraft sind, während K_4 die mit $\frac{i}{c}$ multiplizierte pro Volumen und Zeiteinheit vom Felde auf die Träger der elektrischen Massen abgegebene Energie bedeutet.

Die Allgemeingültigkeit der Erhaltungssätze und des bereits im § 14 abgeleiteten Satzes von der Trägheit der Energie legen es nahe, den Beziehungen (69), (70), (71) und (48a), ~~eine allgemeine~~ obwohl sie an einem ganz speziellen Falle gefunden sind, eine allgemeine Bedeutung zuzuschreiben. Wir verdanken diese Verallgemeinerung, welche den wichtigsten neueren Fortschritt der Relativitätstheorie bildet, den Untersuchungen von Minkowski, Abraham, Planck und Laue. Wir haben jeder Art studierter materieller Vorgänge einen symmetrischen Tensor $T_{\mu\nu}$ zuzuordnen, dessen Komponenten die in (69) angegebene physikalische Bedeutung haben. Gleichung (48a) muss dann stets erfüllt sein. Die zu lösende Aufgabe besteht jeweilen darin, herauszufinden, wie $(T_{\mu\nu})$ aus den die studierten Vorgänge charakterisierenden Variabeln zu bilden ist. Finden in demselben Raume mehrere Vorgänge statt, ~~deren jeder von~~ die in energetischer Beziehung auseinandergehalten werden können, so haben wir jedem solchen besonderen Vorgang seinen eigenen Spannungs-Energie-Tensor $(T^{(1)}_{\mu\nu})$ etc. zuzuschreiben, und $(T_{\mu\nu})$ der Summe dieser Einzeltensoren gleichzusetzen.

Der Spannungs-Energie-Tensor der elektromagnetischen Vorgänge.

Die im letzten § untersuchten elektromagnetischen Zustände setzen sich aus dreierlei Zuständen zusammen, die nach unserem heutigen Wissen energetisch durchaus trennbar sind. Nämlich aus

1) den elektromagnetischen Zuständen des Vakuums
2) den Zuständen der elektrischen Polarisation der Materie
3) den Zuständen der magnetischen Polarisation der Materie.

Diesen ordnen wir der Reihe nach die Spannungs-Energie-Tensoren $(T^0_{\mu\nu}), (T^e_{\mu\nu})$ und $(T^m_{\mu\nu})$ zu.

$(T^0_{\mu\nu})$ hat denselben Ausdruck wie in der Elektrodynamik des Vakuums. Man hat also gemäss (48b)

$$(T^0_{\mu\nu}) = \frac{1}{2}\left\{(\mathfrak{F}_{\mu\sigma})(\mathfrak{F}_{\nu\sigma}) - (\mathfrak{F}^*_{\mu\sigma})(\mathfrak{F}^*_{\nu\sigma})\right\} \quad \ldots \ldots (72)$$

Der Tensor $T^{(e)}_{\mu\nu}$ ist früher für den Fall der Ruhe $(q=0)$ bestimmt worden. Es ergeben sich die p^e_{xx} bis p^e_{zz} für diesen Fall aus (10a). Ferner muss für den Fall der Ruhe $\mathfrak{f}^e = \mathfrak{g}^e = 0$, $W^e = \frac{1}{\varepsilon-1}\frac{\mathfrak{P}^2}{2}$ sein. Hieraus ergibt sich $(T^e_{\mu\nu})$ eindeutig für den Fall der Bewegung zu

$$(T^e_{\mu\nu}) = \frac{1}{\varepsilon-1}\left\{(\mathfrak{P}_{\mu\sigma})(\mathfrak{P}_{\nu\sigma}) - \frac{1}{4}(\delta_{\mu\nu})(\mathfrak{P}_{\sigma\tau})(\mathfrak{P}_{\sigma\tau})\right\} \quad \ldots (73)$$

Analog muss sein

$$(T^m_{\mu\nu}) = \frac{1}{\mu-1}\left\{(\mathfrak{M}_{\mu\sigma})(\mathfrak{M}_{\nu\sigma}) - \frac{1}{4}(\delta_{\mu\nu})(\mathfrak{M}_{\sigma\tau})(\mathfrak{M}_{\sigma\tau})\right\} \quad \ldots (74)$$

where $(\delta_{\mu\sigma})$ denotes the special symmetric tensor given in §16

$$\begin{matrix} 1 & 0 & 0 & 0 \\ 0 & 1 & 0 & 0 \\ 0 & 0 & 1 & 0 \\ 0 & 0 & 0 & 1 \end{matrix}$$

Here

$$(T_{\mu\nu}) = (T^0_{\mu\nu}) + (T^e_{\mu\nu}) + (T^m_{\mu\nu}) \qquad \ldots (75)$$

If we express the components of $(T^e_{\mu\nu})$ in the customary vectorial notation, we obtain

$$\left.\begin{aligned} p^e_{xx} &= \frac{1}{\varepsilon - 1}\{-\mathfrak{p}_x^{*2} - \mathfrak{p}_x^{**2} + \frac{1}{2}(\mathfrak{p}^{*2} + \mathfrak{p}^{**2})\} \\ p_{xy} &= \frac{1}{\varepsilon - 1}\{-\mathfrak{p}^*_x\mathfrak{p}^*_y - \mathfrak{p}^{**}_x\mathfrak{p}^{**}_y\} \\ &\text{etc.} \\ \mathbf{s}^e &= c^2\mathfrak{g}^e = \frac{c}{\varepsilon - 1}[\mathfrak{p}^{**}, \mathfrak{p}^*] \\ \eta^e &= \frac{1}{\varepsilon - 1}\frac{\mathfrak{p}^{*2} + \mathfrak{p}^{**2}}{2} \end{aligned}\right\} \quad (76)$$

The expressions for the corresponding components of $(T^m_{\mu\nu})$ are completely analogous. Let us also write down the components of $T^0_{\mu\nu}$ in vectorial notation.

$$\left.\begin{aligned} p^0_{xx} &= -\mathfrak{e}_x^2 - \mathfrak{h}_x^2 + \frac{1}{2}(\mathfrak{e}^2 + \mathfrak{h}^2) \\ p^0_{xy} &= -\mathfrak{e}_x\mathfrak{e}_y - \mathfrak{h}_x\mathfrak{h}_y \\ &\text{etc.} \\ \mathbf{s}^0 &= c^2\mathfrak{g}^0 = c[\mathfrak{e}, \mathfrak{h}] \\ \eta^0 &= \frac{\mathfrak{e}^2 + \mathfrak{h}^2}{2} \end{aligned}\right\} \quad (76a)$$

If we now assume that the four-vector (K_μ) of the force density, whose first three components constitute the customary vector $\mathfrak{f}$ of the force density, is formed according to (48a), it will decompose again into three summands, (K^0_μ), (K^e_μ), and (K^m_μ). Of these, we are especially interested in the first part (K^0_μ), because its expression is independent of the assumed law that connects polarizations and field strengths. Thus, we form the divergence of $(T^0_{\mu\nu})$ and rewrite it with the help of differential equations (Id). We obtain then

$$\left.\begin{aligned} \mathfrak{f}^0 &= -\mathfrak{e}\,\mathrm{div}\,\mathfrak{p}^{**} - \mathfrak{h}\,\mathrm{div}\,\mathfrak{m}^{**} + \mathfrak{e}\{\frac{1}{c}(\mathfrak{i}, \frac{\mathfrak{q}}{c}) + \rho\} + \left[\frac{\mathfrak{i}}{c} + \frac{\mathfrak{q}}{c}\rho, \mathfrak{h}\right] \\ &\quad + \left[\frac{\dot{\mathfrak{p}}^{**}}{c} - \mathrm{rot}\,\mathfrak{p}^*, \mathfrak{h}\right] - \left[\frac{\dot{\mathfrak{m}}^{**}}{c} - \mathrm{rot}\,\mathfrak{m}^*, \mathfrak{e}\right] \\ \varphi^0 &= \mathfrak{e}(\mathfrak{i} + \rho\mathfrak{q}) + c\left(\mathfrak{e}, \frac{\dot{\mathfrak{p}}^{**}}{c} - \mathrm{rot}\,\mathfrak{p}^*\right) + c\,(\mathfrak{h}, \frac{\dot{\mathfrak{m}}^{**}}{c} - \mathrm{rot}\,\mathfrak{m}^*) \end{aligned}\right\} \quad (77)$$

$\varphi_0 = K^0_4 \cdot \frac{c}{i}$ denotes the energy transferred to the matter, which corresponds to the tensor $(T^0_{\mu\nu})$. In the case where the matter under consideration is electrically and magnetically rigid, i.e., the polarizations of the material particles cannot change, (77)

$(\delta_{\mu\sigma})$ bedeutet dabei den in § 16 angegebenen speziellen symmetrischen Tensor

$$\begin{matrix} 1 & 0 & 0 & 0 \\ 0 & 1 & 0 & 0 \\ 0 & 0 & 1 & 0 \\ 0 & 0 & 0 & 1 \end{matrix} .$$

Dabei ist

$$(T_{\mu\nu}) = (T_{\mu\nu}^{0}) + (T_{\mu\nu}^{e}) + (T_{\mu\nu}^{m}) \quad \ldots (25)$$

Drücken wir die Komponenten von $(T_{\mu\nu}^{e})$ in gewöhnlicher Vektorschreibweise aus, so erhalten wir

$$\left.\begin{aligned} p_{xx}^{e} &= \frac{1}{\varepsilon-1}\left\{-\mathfrak{e}_x^{*2} - \mathfrak{h}_x^{**2} + \frac{1}{2}(\mathfrak{e}^{*2}+\mathfrak{h}^{**2})\right\} \\ p_{xy} &= \frac{1}{\varepsilon-1}\left\{-\mathfrak{e}_x^{*}\mathfrak{e}_y^{*} - \mathfrak{h}_x^{**}\mathfrak{h}_y^{**}\right\} \\ &\text{etc.} \\ \mathfrak{f}^{e} &= c^2\mathfrak{g}^{e} = \frac{c}{\varepsilon-1}\left[\mathfrak{e}^{**}, \mathfrak{h}^{*}\right] \\ \eta^{e} &= \frac{1}{\varepsilon-1}\,\frac{\mathfrak{e}^{*2}+\mathfrak{h}^{**2}}{2} \end{aligned}\right\} \quad (26)$$

Ganz analog sind die Ausdrücke für die entsprechenden Komponenten von $(T_{\mu\nu}^{m})$. Wir schreiben noch die Komponenten von $T_{\mu\nu}^{0}$ in Vektorschreibweise hin.

$$\left.\begin{aligned} p_{xx}^{0} &= -\mathfrak{e}_x^{2} - \mathfrak{h}_x^{2} + \frac{1}{2}(\mathfrak{e}^{2}+\mathfrak{h}^{2}) \\ p_{xy}^{0} &= -\mathfrak{e}_x\mathfrak{e}_y - \mathfrak{h}_x\mathfrak{h}_y \\ &\text{etc.} \\ \mathfrak{f}^{0} &= c^2\mathfrak{g}^{0} = c[\mathfrak{e}, \mathfrak{h}] \\ \eta^{0} &= \frac{\mathfrak{e}^{2}+\mathfrak{h}^{2}}{2} \end{aligned}\right\} \quad (26a)$$

Denken wir uns nun gemäss (48a) den Vierervektor (K_μ) der Kraftdichte gebildet, dessen drei erste Komponenten den gewöhnlichen Vektor $\mathfrak{f}$ der Kraftdichte bilden, so zerfällt dieser wieder in drei Summanden (K_μ^{0}), (K_μ^{e}) und (K_μ^{m}). Hievon interessiert uns besonders der erste Teil (K_μ^{0}), denn ~~er~~ sein Ausdruck ist von dem angenommenen Gesetze unabhängig, nach welchem Polarisationen und Feldstärken verknüpft sind. Wir bilden also die Divergenz von $(T_{\mu\nu}^{0})$ und formen sie mit Hilfe der Differentialgleichungen (Id) um. Wir erhalten dann

$$\left.\begin{aligned} \mathfrak{f}^{0} &= -\mathfrak{e}\,\operatorname{div}\mathfrak{p}^{**} - \mathfrak{h}\,\operatorname{div}\mathfrak{m}^{**} + \mathfrak{e}\left\{\frac{1}{c}\left(\mathfrak{i}, \frac{\mathfrak{q}}{c}\right) + \varrho\right\} + \left[\frac{\mathfrak{i}}{c} + \frac{\mathfrak{q}}{c}\varrho, \mathfrak{h}\right] \\ &\quad \sout{+\left[\mathfrak{h}, \frac{\dot{\mathfrak{p}}^{**}}{c} - \operatorname{rot}\mathfrak{p}^{*}\right]} - \left[\mathfrak{e},\right. \\ &\quad + \left[\frac{\dot{\mathfrak{p}}^{**}}{c} - \operatorname{rot}\mathfrak{p}^{*}, \mathfrak{h}\right] - \left[\frac{\dot{\mathfrak{m}}^{**}}{c} - \operatorname{rot}\mathfrak{m}^{*}, \mathfrak{e}\right] \\ \varphi^{0} &= \mathfrak{e}(\mathfrak{i} + \varrho\mathfrak{q}) + c\left(\mathfrak{e}, \frac{\dot{\mathfrak{p}}^{**}}{c} - \operatorname{rot}\mathfrak{p}^{*}\right) + c\left(\mathfrak{h}, \frac{\dot{\mathfrak{m}}^{**}}{c} - \operatorname{rot}\mathfrak{m}^{*}\right) \end{aligned}\right\} \quad (27)$$

~~$\frac{i}{c}\varphi_0 = (K_4^{0})$~~ $\frac{i}{c}\varphi_0 = K_4^{0} \cdot \frac{c}{i}$ bedeutet die an die Materie abgegebene Energie, die dem ~~Tensor~~ $(T_{\mu\nu}^{0})$ entspricht. In dem Falle, dass die betrachtete Materie elektrisch ~~und~~ magnetisch starr ist, d. h. dass die Polarisationen der materiellen Teilchen sich nicht ändern können, enthält (27) die Wirkungen auf die Materie vollständig, da dann die Kräfte

contains completely the effects on the matter, for in that case the forces between the matter and the polarization charges do not play any role as regards the energy, and thus the same also applies to the case when the polarization states of the matter do not change temporally relative to the latter. In all other cases, all of those force effects and releases of energy that correspond to the tensors $(T^{e}_{\mu\nu})$ and $(T^{m}_{\mu\nu})$ of polarization also come in. But since the expressions for the latter seem not very edifying and are probably of little physical value, we will not write them down. We restrict ourselves to the treatment of two special cases.

1. The medium is neither electrically nor magnetically polarizable.

For this case we deduce immediately from (77)

$$\left.\begin{aligned} \mathfrak{f} &= \mathfrak{e}\left\{\frac{1}{c}\left(\mathfrak{i}, \frac{\mathfrak{q}}{c}\right) + \rho\right\} + \left[\frac{\mathfrak{i}}{c} + \frac{\mathfrak{q}}{c}\rho, \mathfrak{h}\right] \\ \varphi &= \mathfrak{e}\,(\mathfrak{i} + \rho\mathfrak{q}) \end{aligned}\right\} \qquad (77a)$$

φ is composed of the work transferred to the matter per unit volume and unit time $(\mathfrak{f}, \mathfrak{q})$, and of the heat w, so that we have*

$$w = \phi - (\mathfrak{f}, \mathfrak{q})$$

or, if one substitutes the values from (77a) for $\mathfrak{f}$ and φ and uses (66),

$$w = \sigma\left\{\mathfrak{e}^{*2} - \left(\frac{\mathfrak{q}}{c}, \mathfrak{e}^{*}\right)^{2}\right\}\sqrt{1 - \frac{q^2}{c^2}} \qquad \ldots (78a)$$

2. The case of rest ($\mathfrak{q} = 0$).

Here the expressions are simplified significantly. The expression for $\mathfrak{f}$ for this case has already been given in formula (11a) of §3.

§21. Inertia of Energy. Dynamics of the Mass Point

The Integral Form of the Conservation Laws. First we will show that equation (48a) really includes the conservation laws and the law of the inertia of energy. Let us consider a spatially bounded system that is under the influence of volume forces. The system forms a four-dimensional thread in the four-dimensional manifold.[90] We mark off a piece of it that is cut out by the time variables x'_4 and x''_4, and integrate equation (48a) over this volume. In that way we obtain for the individual component

$$-\int K_\mu dx_1 \ldots dx_4 = \int \sum \frac{\partial T_{\mu\nu}}{\partial x_\nu} dx_1 \ldots dx_4 = \int \frac{\partial T_{\mu 4}}{\partial x_4} dx_1 \ldots dx_4.$$

The last reformulation is based on the circumstance that the $T_{\mu\nu}$ vanish at the spatial integration limits. We can bring the integral on the right side into the form

$$\int dx_4 \frac{\partial}{\partial x_4}\left\{\int T_{\mu 4} dx_1 dx_2 dx_3\right\}.$$

If we introduce the notations

---

* In accordance with (78), one can also write $w = -c\,(K_\mu)\,(G_\mu)\sqrt{1 - \frac{q^2}{c^2}}$. Thus, $\frac{w}{\sqrt{1 - \frac{q^2}{c^2}}}$ is a scalar, and is equal to the the quantity w_0 if the latter denotes the heat developed per unit volume and time. Integrating over a four-dimensional volume in which $\mathfrak{q}$ = const., one also obtains for the total quantity of heat w

$$w = w_0\sqrt{1 - \frac{q^2}{c^2}}.$$

zwischen Materie und Polarisationsladungen energetisch keine Rolle spielen, ebenso also auch in dem Falle, dass die Polarisationszustände der Materie relativ zu letzterer sich zeitlich nicht ändern. In allen anderen Fällen kommen noch diejenigen Kraftwirkungen und Energieabgaben hinzu, welche den Tensoren $(T^e_{\mu\nu})$ und $(T^m_{\mu\nu})$ der Polarisation entsprechen. Da die Ausdrücke für jene aber wenig anschaulich aussehen und wohl auch kaum grossen physikalischen Wert haben dürften, wollen wir sie nicht hinschreiben. Wir beschränken uns auf die Behandlung zweier Spezialfälle.

1. ~~Ist~~ das Medium ist weder elektrisch noch magnetisch polarisierbar.

Für diesen Fall entnimmt man (27) unmittelbar

$$\left.\begin{aligned} \mathfrak{f} &= \mathfrak{e}\left\{\frac{1}{c}\left(\mathfrak{i},\frac{\mathfrak{q}}{c}\right)+\rho\right\} + \left[\frac{\mathfrak{i}}{c}+\frac{\mathfrak{q}}{c}\rho,\mathfrak{h}\right] \\ \varphi &= \mathfrak{e}(\mathfrak{i}+\rho\mathfrak{q}) \end{aligned}\right\} \quad (27a)$$

φ setzt sich zusammen aus der pro Volumeneinheit und Zeiteinheit auf die Materie übertragenen Arbeit $(\mathfrak{f},\mathfrak{q})$ und Wärme w, sodass man hat*

$$w = \varphi - (\mathfrak{f},\mathfrak{q}), \quad \ldots (28)$$

oder, indem man für $\mathfrak{f}$ und φ aus (27a) die Werte einsetzt, und unter Benutzung von (66)

$$W = \sigma\left\{\mathfrak{e}^{\times 2} - \left(\frac{\mathfrak{q}}{c},\mathfrak{e}^{\times}\right)^2\right\}\sqrt{1-\frac{q^2}{c^2}} \quad \ldots (28a)$$

~~$W = \sigma\,\mathfrak{e}^2\sqrt{1-\frac{q^2}{c^2}} \ldots (28a)$~~

2.) Fall der Ruhe $(\mathfrak{q} = 0)$.

Hier ~~reduzieren~~ sich die Ausdrücke bedeutend / der Ausdruck für $\mathfrak{f}$ ~~man erhält für $\mathfrak{f}$~~ für diesen Fall ist in Formel (11a) des § 3 bereits angegeben.

~~$$\left.\begin{aligned} \mathfrak{f} &= \mathfrak{e}\rho - \mathrm{grad}\left(\varepsilon\frac{\mathfrak{e}^2}{2}\right) - \mathrm{grad}\left(\mu\frac{\mathfrak{h}^2}{2}\right) - [\mathfrak{p},\mathrm{rot}\,\mathfrak{e}] - [\mathfrak{m},\mathrm{rot}\,\mathfrak{h}] \\ &+ \frac{1}{c}[\mathfrak{i}+\dot{\mathfrak{p}},\mathfrak{h}] - \frac{1}{c}[\dot{\mathfrak{m}},\mathfrak{e}] \end{aligned}\right\} \quad (29)$$~~

§ 21. Trägheit der Energie. ~~Dynamik des Massenpunktes.~~

Integralform der Erhaltungssätze.

Wir wollen zunächst zeigen, dass ~~die Existenz einer~~ ~~Gültigkeit der~~ Gleichung ~~von der Gestalt~~ (48a) wirklich die ~~I~~ Erhaltungssätze und den Satz von der Trägheit der Energie in sich schliesst. Es werde ein räumlich begrenztes System, das unter dem Einfluss von Volumkräften stehe, betrachtet. Dasselbe bildet in der vierdimensionalen Mannigfaltigkeit einen vierdimensionalen Faden; von diesem grenzen wir ein Stück ab, das durch die Zeitvariabeln x_4' und x_4'' herausgeschnitten wird, und ~~bilden das vierfache Vol~~ integrieren die Gleichung (48a) über dies Volumen. Wir erhalten so für die einzelne Komponente

~~$$\int (K_\mu)\,dV = \int\left(\frac{\partial T_{\mu\nu}}{\partial x_\nu}\right)\left(\sum\frac{\partial T_{\mu\nu}}{\partial x_\nu}\right)dV = \int\frac{\partial_{\mu 4}}{\partial x_4}$$~~

$$-\int K_\mu\,dx_1\cdots dx_4 = \int\left(\sum\frac{\partial T_{\mu\nu}}{\partial x_\nu}\right)dx_1\cdots dx_4 = \int\frac{\partial T_{\mu 4}}{\partial x_4}\,dx_1\cdots dx_4.$$

Die letzte Umformung beruht darauf, dass an den räumlichen Integrationsgrenzen die $T_{\mu\nu}$ verschwinden. ~~Indem~~ Das Integral rechts können wir in die Form

$$\int dx_4\frac{\partial}{\partial x_4}\left\{\int T_{\mu 4}\,dx_1\,dx_2\,dx_3\right\}$$

bringen. Führen wir noch die Bezeichnungen ein

---

* Nach (28) kann man auch schreiben $w = -(K_\mu)(G_\mu)\sqrt{1-\frac{q^2}{c^2}}$. $\frac{w}{\sqrt{1-\frac{q^2}{c^2}}}$ ist also ein Skalar, und zwar gleich der Grösse w_0, wenn letzteres die von einem mitbewegten Beobachter aus beurteilte Wärmeentwicklung pro Volumen- und Zeiteinheit bedeutet. Integriert man über ein vierdimensionales Volumen, in welchem $\mathfrak{q}$ = konst. ist, so erhält man für die gesamten Wärmemengen W ebenfalls die Beziehung

$$W = W_0\sqrt{1-\frac{q^2}{c^2}}.$$

$$\int K_\mu dx_1 dx_2 dx_3 = \int K_\mu dxdydz = \overline{K_\mu}$$

$$\int T_{\mu 4} dxdydz = \overline{T_{\mu 4}},$$

we obtain the equations

$$\left.\begin{aligned} -\int \overline{K}_\mu dx_4 &= \left|\overline{T}_{\mu 4}\right|_{x'_4}^{x''_4} \\ \text{or} \qquad -\overline{K}_\mu &= \frac{d\overline{T}_{\mu 4}}{dx_4} \end{aligned}\right\} \tag{80}$$

These equations express the conservation laws of momentum and energy in the integral form. We obtain them in the familiar real form by introducing the notations

$$\overline{K}_1 = \bar{\mathfrak{f}}_x = \int \mathfrak{f}_x dxdydz \quad \text{etc} \qquad \Big| \qquad \overline{T}_{14} = ic\bar{\mathfrak{g}}_1 = ic\int \mathfrak{g}_x dxdydz$$

$$\overline{K}_4 = \frac{i}{c}\bar{\varphi} = \frac{i}{c}\int \varphi dxdydz \qquad \Big| \qquad \overline{T}_{44} = -\bar{\eta} = -\int \eta dxdydz$$

Thus, we denote the vector sum of all of the forces acting on the system at one time by $\bar{\mathfrak{f}}$, the quantities of energy transferred to the system per unit time at one point of time by $\bar{\varphi}$, the momentum by $\bar{\mathfrak{g}}$, and the energy of the system by $\bar{\eta}$. We obtain

$$\left.\begin{aligned} \bar{\mathfrak{f}} &= \frac{d\bar{\mathfrak{g}}}{dt} \\ \bar{\varphi} &= \frac{d\bar{\eta}}{dt} \end{aligned}\right\} \tag{80a}$$

These are the conservation laws in the customary form.

We will now prove that $\bar{\mathfrak{g}}_x, \bar{\mathfrak{g}}_y, \bar{\mathfrak{g}}_z, \frac{i}{c}\bar{\eta}$ form a four-vector. K_μ are the components of a four-vector, and thus the integrals

$$\int K_\mu dx_1 dx_2 dx_3 dx_4$$

are also such if the four-dimensional integration limits are defined in a manner independent of the choice of the reference system. To be sure, this does not obtain in our case, since we marked off our four-dimensional thread using the reference-system-dependent condition that x_4 lie between specific limits. But if, up to sufficient distances from these limits, (K_μ) vanishes, or if we can view the thread as infinitely thin, then the change of the integration limits that takes place when the reference system changes does not have any influence on the integral. Then

$$\left|\overline{T}_{\mu 4}\right|_{x_4'}^{x_4''},$$

and thus the increase in $\overline{T}_{\mu 4}$, is a four-vector. If this holds for the increases, then it holds for the $\overline{T}_{\mu 4}$ as well,* from which the assertion follows, with the indicated limitations.

* To be sure, this is not rigorous, because additive constants might be present that do not have the character of a vector; but this seems so artificial that we will not dwell on this possibility at all.

$$\int K_\mu \, dx_1 dx_2 dx_3 = \int K_\mu \, dx\, dy\, dz = \overline{K_\mu}$$

$$\int T_{\mu 4} \, dx\, dy\, dz = \overline{T_{\mu 4}},$$

so erhalten wir die Gleichungen

$$\left.\begin{aligned} -\int \overline{K_\mu}\, dx_4 &= \left|\overline{T_{\mu 4}}\right|_{x_4'}^{x_4''} \\ \text{oder} \qquad -\overline{K_\mu} &= \frac{d\overline{T_{\mu 4}}}{dx_4} \end{aligned}\right\} \quad (80)$$

Diese Gleichungen drücken die Erhaltungssätze des Impulses und der Energie in der Integralform aus. Wir erhalten sie in der geläufigen reellen Form, indem wir die Bezeichnungen einführen

$$\overline{K_1} = \overline{f_x} = \int f_x \, dx\, dy\, dz \quad \text{etc} \qquad \Bigg| \qquad \overline{T_{14}} = ic\,\overline{y_1} = ic \int y_x \, dx\, dy\, dz$$

$$\overline{K_4} = \frac{i}{c}\overline{\varphi} = \frac{i}{c}\int \varphi \, dx\, dy\, dz \qquad \Bigg| \qquad \overline{T_{44}} = -\overline{\eta} = -\int \eta \, dx\, dy\, dz.$$

Wir bezeichnen also ~~mit~~ $\overline{f}$ die Vektorsumme aller auf das System (zu einer Zeit) wirkenden Kräfte, mit $\overline{\varphi}$ die (Summe der) zu einer Zeit pro Zeiteinheit auf das System übertragenen ~~Arbeiten~~ Energiemengen, (mit $\overline{y}$ den Impuls mit $\overline{\eta}$ die Energie des Systems.) Es ergibt sich

$$\left.\begin{aligned} \overline{f} &= \frac{d\overline{y}}{dt} \\ \overline{\varphi} &= \frac{d\overline{\eta}}{dt} \end{aligned}\right\} \quad (80a)$$

Dies sind die Erhaltungssätze in gewöhnlicher Form.

Wir werden nun beweisen, dass $\overline{y_x}, \overline{y_y}, \overline{y_z}, \frac{i}{c}\overline{\eta}$ einen Vierervektor bilden. K_μ sind die Komponenten eines Vierervektors, also auch die Integrale

$$\int K_\mu \, dx_1 dx_2 dx_3 dx_4,$$

falls die (vierdimensionalen) Integrationsgrenzen in einer von der Wahl des Bezugssystems unabhängigen Weise definiert sind. Dies ist nun allerdings in unserem Falle nicht zutreffend, da wir unseren vierdimensionalen Faden durch die vom Bezugssystem abhängige Bedingung abgegrenzt haben, dass x_4 zwischen bestimmten Grenzen liegen. Wenn aber ~~in~~ bis zu hinreichenden Abständen von diesen Grenzen (K_μ) verschwindet, oder wenn wir den Faden als unendlich dünn ansehen dürfen, hat die bei Änderung des Bezugssystems auftretende Änderung der Integrationsgrenzen keinen Einfluss auf das Integral. Dann ist

$$\left|\overline{T_{\mu 4}}\right|_{x_4'}^{x_4''},$$

also der Zuwachs von $\overline{T_{\mu 4}}$ ein Vierervektor. Wenn dies für die Zuwächse gilt, so gilt es für die $\overline{T_{\mu 4}}$ ebenfalls,* woraus (mit den angegebenen Beschränkungen) die Behauptung folgt.

Impuls und Energie des Massenpunktes

~~Der Massenpunkt der Mechanik ist durch einen Skalar m charakterisiert. Sein Vierervektor des Impulses und der Energie wird ausserdem vom Vierervektor der Geschwindigkeit (G_μ) abhängen; Er muss daher~~

~~$m\,(G_\mu)$~~

---

* Dies ist allerdings nicht streng, da additive Konstante vorhanden sein könnten, denen der Vektorcharakter abgeht; dies erscheint aber so künstlich, dass wir uns mit dieser Möglichkeit gar nicht weiter beschäftigen.

Inertia of Energy

We consider now a closed system from various systems of reference and assume that it is possible to choose a system of reference in such a way that the momentum $\bar{\mathfrak{g}}$ of the system under consideration vanishes relative to the system of reference. In that case the four-vector

$$\bar{\mathfrak{g}}_x \quad \bar{\mathfrak{g}}_y \quad \bar{\mathfrak{g}}_z \quad \frac{i}{c}\bar{\eta}$$

of the momentum and of the energy degenerates to

$$0 \quad 0 \quad 0 \quad \frac{i}{c}\bar{\eta}$$

But in the special case where $\mathfrak{q} = 0$,[1] the latter system of components belongs to the four-vector

$$\frac{\bar{\eta}_0}{c}(G_\mu).$$

Consequently, the latter four-vector is identical to the four-vector of the momentum and energy for any other arbitrary systems as well. By equating we obtain the equations

$$\left.\begin{aligned} \bar{\mathfrak{g}} &= \frac{\bar{\eta}_0}{c^2}\frac{\mathfrak{q}_x}{\sqrt{1-\frac{q^2}{c^2}}} \\ \bar{\eta} &= \bar{\eta}_0\frac{1}{\sqrt{1-\frac{q^2}{c^2}}} \end{aligned}\right\} \tag{81}$$

or to a first degree of approximation

$$\left.\begin{aligned} \bar{\mathfrak{g}} &= \frac{\bar{\eta}_0}{c^2}\mathfrak{q} \\ \bar{\eta} &= \bar{\eta}_0 + \frac{\bar{\eta}_0}{c^2}\frac{q^2}{2} \end{aligned}\right\} \tag{81a}$$

From these equations it follows that $\frac{\bar{\eta}_0}{c^2}$ plays the role of the total mass M of the system in the sense of ordinary mechanics. The energy and total mass (of a closed system) differ from each other only by a universal factor, and thus are completely equivalent. The laws of conservation of mass and of conservation of energy fuse into a single law. (cf. §14)

Equation of Motion of the Mass Point

We obtain a mass point in the sense of mechanics if we consider a system that is adequately characterized from the standpoint of a comoving observer by its "rest-mass" $\frac{\bar{\eta}_0}{c^2} = m$, where this rest-mass may be viewed as unchanging. According to (81), the momentum and energy take the form

$$\left.\begin{aligned} \bar{\mathfrak{g}} &= \frac{m\mathfrak{q}}{\sqrt{1-\frac{q^2}{c^2}}} \\ \bar{\eta} &= \frac{mc^2}{\sqrt{1-\frac{q^2}{c^2}}} \end{aligned}\right\} \tag{81b}$$

[1] Here $\bar{\eta}_0$ is to be viewed as a scalar, which is admissible in accordance with the definition of this quantity rest-energy).

Trägheit der Energie.

Wir betrachten nun ein abgeschlossenes System von verschiedenen Bezugssystemen aus und nehmen an, dass es möglich sei, ein Bezugssystem so zu wählen, dass der Impuls $\bar{\mathfrak{J}}$ des ~~Bezugssystems~~ betrachteten ~~Bezugs~~Systems relativ zum Bezugssystem verschwinde. Der Vierervektor

$$\bar{\mathfrak{J}}_x \quad \bar{\mathfrak{J}}_y \quad \bar{\mathfrak{J}}_z \quad \frac{i}{c}\bar{\eta}$$

des Impulses und der Energie degeneriert dann in

$$0 \quad 0 \quad 0 \quad \frac{i}{c}\bar{\eta}_0 .$$

Das letztere Komponentensystem gehört aber zu dem Vierervektor

$$\frac{\bar{\eta}_0}{c}(\mathfrak{G}_\nu)$$

in dem Spezialfall $\mathfrak{q} = 0$.[1] Folglich ist der letztere Vierervektor auch für beliebige andere Bezugssysteme mit dem Vierervektor des Impulses und der Energie identisch. Man erhält durch Gleichsetzen die Gleichungen

$$\left.\begin{aligned} \bar{\mathfrak{J}} &= \frac{\bar{\eta}_0}{c^2}\frac{\mathfrak{q}_x}{\sqrt{1-\frac{\mathfrak{q}^2}{c^2}}} \\ \bar{\eta} &= \bar{\eta}_0 \frac{1}{\sqrt{1-\frac{\mathfrak{q}^2}{c^2}}} \end{aligned}\right\} \quad (81)$$

oder in erster Näherung

$$\left.\begin{aligned} \bar{\mathfrak{J}} &= \frac{\bar{\eta}_0}{c^2}\mathfrak{q} \\ \bar{\eta} &= \bar{\eta}_0 + \frac{\bar{\eta}_0}{c^2}\frac{\mathfrak{q}^2}{2} \end{aligned}\right\} \quad (81a)$$

Aus diesen Gleichungen geht hervor, dass $\frac{\bar{\eta}_0}{c^2}$ die Rolle der Gesamtmasse M des Systems im Sinne der gewöhnlichen Mechanik spielt. Energie und Gesamtmasse (eines abgeschlossenen Systems) unterscheiden sich nur durch einen universellen Faktor, sind also vollkommen äquivalent. Die Sätze von der Erhaltung der Masse und der Erhaltung der Energie verschmelzen in einen einzigen (vgl. § 14).

Bewegungsgleichung des Massenpunktes.

Einen Massenpunkt im Sinne der Mechanik erhalten wir, indem wir ein System betrachten, das vom Standpunkt eines mitbewegten Beobachters durch seine „Ruhemasse" $\frac{\eta_0}{c^2} = m$ genügend charakterisiert ist, wobei diese Ruhemasse als unveränderlich betrachtet werden darf. Impuls und Energie erhalten (gemäss (81)) die Form

$$\left.\begin{aligned} \bar{\mathfrak{J}} &= \frac{m\mathfrak{q}}{\sqrt{1-\frac{\mathfrak{q}^2}{c^2}}} \\ \bar{\eta} &= \frac{mc^2}{\sqrt{1-\frac{\mathfrak{q}^2}{c^2}}} \end{aligned}\right\} \quad (81b)$$

---

1) $\bar{\eta}_0$ ist hiebei als Skalar aufzufassen, was ~~seiner~~ nach der Definition dieser Grösse zulässig ist (Ruhe-Energie)

and the first of equations (80a) yields the law of motion

$$\mathfrak{f} = \frac{d}{dt}\left\{\frac{m\mathfrak{q}}{\sqrt{1-\frac{q^2}{c^2}}}\right\}. \qquad \ldots (82)$$

If the mass point possesses a charge e and is moving under the sole influence of an electromagnetic field, then we have to set, according to ( ), [91]

$$\mathfrak{f} = e\left\{\mathfrak{e} + \left[\frac{\mathfrak{q}}{c}, \mathfrak{h}\right]\right.$$

One thus obtains the familiar laws of motion of the electron for quasi-stationary motion.

§22. Euler's Hydrodynamical Equations from the Standpoint of the Relativity Theory [92]

In addition, we will discuss the equations for the (adiabatic) motion of an incompressible liquid as an example of the application of the conservation laws. The path to be taken is the following one. First we find easily the stress-energy tensor for the *stationary* liquid. From this we obtain the expression for the same tensor when referred to an arbitrary justified reference system. Then equation (48a) yields directly the equations we are seeking, including the continuity equation.

Let the stationary liquid be characterized by the pressure p and the density μ, which shall be given as a function of p. We define the density μ in the sense of the results of the last §, by postulating that μc^2 is the (total!) energy of the liquid per unit volume. Next we have to investigate how μ is determined by the equation of state of the liquid.

Imagine that a given quantity of the liquid is enclosed in a (massless) shell of a changeable volume V at the pressure p. V is then given as a function of p. In the sense of the previous §, the liquid and the shell together form a closed system of inertial mass m, which is a function of the volume. For if an initial state of the system is characterized by the values p_1, V_1, m_1, then, according to the results of the previous §, we will have for an arbitrary state (p, V, m)

$$(m - m_1)c^2 = -\int_{V_1}^{V} p\,dV,$$

hence

$$m = m_1 - \frac{1}{c^2}\int_{V_1}^{V} p\,dV$$

and

$$\mu = \frac{m}{V} = \frac{m_1}{V} - \frac{1}{c^2 V}\int_{V_1}^{V} p\,dV. \qquad \ldots (83)$$

This defines μ as a function of p if the equation of state is known.—
Stress-energy tensor

und die erste der Gleichungen (80a) liefert das Bewegungsgesetz

$$\bar{f} \; \cancel{(\overline{-f})} = \frac{d}{dt}\left\{\frac{mq}{\sqrt{1-\frac{q^2}{c^2}}}\right\} \quad \ldots\ldots (82)$$

Besitzt der Massenpunkt eine Ladung e, und bewegt er sich unter dem alleinigen Einfluss eines elektromagnetischen Feldes, so ist dabei gemäss (   )

$$\bar{f} \; \cancel{-\bar{f}\,(\overline{-f})} = e\left\{\mathfrak{e} + \left[\frac{q}{c}, \mathfrak{h}\right]\right\}$$

zu setzen. Man erhält so die bekannten Bewegungsgleichungen des Elektrons ~~in~~ für quasistationäre Bewegung.

§ 22. Eulers hydrodynamische Gleichungen vom Standpunkt der Relativitätstheorie.

Wir wollen als Beispiel für die Anwendung der Erhaltungssätze noch die Gleichungen der (adiabatischen) Bewegung einer inkompressiblen Flüssigkeit behandeln. Der dabei einzuschlagende Weg ist folgender. Wir finden zunächst leicht den Spannungs-Energietensor für die ruhende Flüssigkeit. Hieraus folgt der Ausdruck desselben Tensors, ~~vom Standpunkt~~ bezogen auf ein beliebiges berechtigtes Bezugssystem. Gleichung (48a) liefert dann unmittelbar die gesuchten Gleichungen mit Einschluss der Kontinuitätsgleichung.

Die ruhende Flüssigkeit sei durch den Druck p und die Dichte μ charakterisiert, welche in Funktion von p gegeben sei. Die ~~Dichte~~ μ definieren wir im Sinne der Ergebnisse des letzten § durch die Festsetzung, dass μc^2 die (gesamte!) Energie der Flüssigkeit pro Volumeneinheit sei. Wir haben zunächst ~~nun~~ zu untersuchen, wie μ durch die Zustandsgleichung der Flüssigkeit bestimmt wird.

Wir denken uns eine gewisse Menge der Flüssigkeit in eine (masselose) Hülle vom veränderbaren Volumen V eingeschlossen beim Drucke p. V ist dann in Funktion von p gegeben. Flüssigkeit nebst Hülle bilden ein abgeschlossenes System im Sinne des vorigen § von der trägen Masse m, die Funktion des Volumens ist. Sei nämlich ein Anfangszustand des Systems durch die Werte p_1, V_1, m_1 charakterisiert, so ist für einen beliebigen Zustand (p, V, m) nach den Ergebnissen des vorigen §

$$(m - m_1)c^2 = -\int_{V_1}^{V} p\,dV,$$

also $$m = m_1 - \frac{1}{c^2}\int_{V_1}^{V} p\,dV$$

und $$\mu = \frac{m}{V} = \frac{m_1}{V} - \frac{1}{c^2 V}\int_{V_1}^{V} p\,dV \quad \ldots\ldots (83)$$

Damit ist μ als Funktion von p definiert, wenn die Zustandsgleichung bekannt ist. —

Spannungs-Energie-Tensor.

Für den Fall, dass das Bezugssystem relativ zur Flüssigkeit ruht, ist —

For the case where the reference system is at rest relative to the liquid, the tensor $(T_{\nu\rho})$ is given—as a glance at (69) will show—by the following system of values:

$$\begin{matrix} p & 0 & 0 & 0 \\ 0 & p & 0 & 0 \\ 0 & 0 & p & 0 \\ 0 & 0 & 0 & -\mu c^2 \end{matrix}$$

Here p and μ are, by their definitions, scalars, i.e., they are defined independently of the <coordinate> reference system. With this in mind, one realizes that

$$(T_{\nu\rho}) = p(\delta_{\nu\rho}) + (\mu c^2 + p)(G_\nu)(G_\rho) \qquad \ldots (84)$$

is a tensor the components of which assume the values just indicated in the special case where $\mathfrak{q} = 0$. Thus, $(T_{\nu\rho})$ is the desired stress-energy tensor of our ideal liquid. With the aid of (84) it is easy to calculate the energy, momentum, etc., of the parts of the system that have a constant density and are under constant pressure, as for example for cavity radiation. However, we will not go into this, but will, instead, derive straight away the equations of motion. According to (48a), they read in the four-dimensional form [93]

$$(-K_\mu) = \left(\sum_\rho \frac{\partial T_{\nu\rho}}{\partial x_\rho}\right) \qquad \ldots (85)$$

or in a more detailed notation <; in the a absence of external forces,> [94]

$$-\mathfrak{f}_x = \frac{\partial p}{\partial x} + \frac{\mu^x \mathfrak{q}_x}{w^2}\frac{\partial \mathfrak{q}_x}{\partial x} + \frac{\mu^x \mathfrak{q}_y}{w^2}\frac{\partial \mathfrak{q}_x}{\partial y} + \frac{\mu^x \mathfrak{q}_z}{w^2}\frac{\partial \mathfrak{q}_x}{\partial z} + \frac{\mu^x}{w^2}\frac{\partial \mathfrak{q}_x}{\partial t}$$

$$+\mathfrak{q}_x \left\{ \frac{\partial}{\partial x}\left(\frac{\mu^x \mathfrak{q}_x}{w^2}\right) + \frac{\partial}{\partial y}\left(\frac{\mu^x \mathfrak{q}_y}{w^2}\right) + \frac{\partial}{\partial z}\left(\frac{\mu^x \mathfrak{q}_z}{w^2}\right) + \frac{\partial}{\partial t}\left(\frac{\mu^x}{w^2}\right) \right\}$$

$$-\frac{i}{c}\eta = -\frac{i}{c}\frac{\partial p}{\partial t}$$

$$-\mathfrak{f}_x = \frac{\partial p}{\partial x} + \frac{\mu^*}{1 - \frac{q^2}{c^2}} \left\{ \mathfrak{q}_x \frac{\partial \mathfrak{q}_x}{\partial x} + \mathfrak{q}_y \frac{\partial \mathfrak{q}_x}{\partial y} + \mathfrak{q}_z \frac{\partial \mathfrak{q}_x}{\partial z} + \frac{\partial \mathfrak{q}_x}{\partial t} \right) + \mathfrak{q}_x A$$

$$-\frac{1}{c^2}\eta = -\frac{1}{c^2}\frac{\partial p}{\partial t} + A,$$

where we have set for brevity

$$A = \frac{\partial}{\partial x}\left(\frac{\mu^x \mathfrak{q}_x}{1 - \frac{q^2}{c^2}}\right) + \frac{\partial}{\partial y}\left(\frac{\mu^x \mathfrak{q}_y}{1 - \frac{q^2}{c^2}}\right) + \frac{\partial}{\partial z}\left(\frac{\mu^x \mathfrak{q}_z}{1 - \frac{q^2}{c^2}}\right) + \frac{\partial}{\partial t}\left(\frac{\mu^x}{1 - \frac{q^2}{c^2}}\right)$$

$$\mu^x = \mu + \frac{p}{c^2}$$

wie ein Blick auf (69) lehrt – der Tensor $(T_{\nu\varrho})$ durch folgendes Wertsystem gegeben

$$\begin{matrix} p & 0 & 0 & 0 \\ 0 & p & 0 & 0 \\ 0 & 0 & p & 0 \\ 0 & 0 & 0 & -\mu c^2 \end{matrix}$$

Dabei sind p und μ ihrer Definition gemäss Skalare, d. h. vom ~~Koordinaten~~ Bezugssystem unabhängig definiert. Mit Rücksicht darauf erkennt man, dass ~~der Tensor~~

$$(T_{\nu\varrho}) = p(\delta_{\nu\varrho}) + (\mu c^2 + p)(\mathfrak{G}_\nu)(\mathfrak{G}_\varrho) \quad \cdots (84)$$

ein Tensor ist, ~~der~~ dessen Komponenten im Spezialfall $\mathfrak{q} = 0$ die eben angegebenen Werte annehmen. $(T_{\nu\varrho})$ ist also der gesuchte Spannungs-Energie-Tensor unserer idealen Flüssigkeit. Mit Hilfe von (84) ist es ein Leichtes, Energie, Impuls etc. von Systemteilen zu berechnen, welche konstante Dichte haben und unter konstantem Drucke stehen, beispielsweise für die Hohlraumstrahlung. Wir wollen jedoch hierauf nicht eingehen, sondern gleich die Bewegungsgleichungen ableiten. Diese ~~sind~~ lauten nach (48a) in vierdimensionaler Form

$$(-K_\mu) = \left(\sum_\varrho \frac{\partial T_{\nu\varrho}}{\partial x_\varrho}\right) \quad \cdots\cdots (85)$$

oder in ~~dreidimensionaler~~ ausführlicher Schreibweise~~, wenn äussere Kräfte fehlen.~~

~~$$-f_x = \frac{\partial p}{\partial x} + \frac{\mu^* \mathfrak{q}_x}{W^2}\frac{\partial}{\partial x}\left(\frac{\mathfrak{q}_x}{W}\right) + \frac{\mu^* \mathfrak{q}_y}{W^2}\frac{\partial}{\partial y}\left(\frac{\mathfrak{q}_x}{W}\right) + \frac{\mu^* \mathfrak{q}_z}{W^2}\frac{\partial}{\partial z}\left(\frac{\mathfrak{q}_x}{W}\right) + \frac{\mu^*}{W^2}\frac{\partial}{\partial t}\left(\frac{\mathfrak{q}_x}{W}\right)$$~~

~~$$+ \frac{\mathfrak{q}_x}{W}\left\{\frac{\partial}{\partial x}\left(\frac{\mu^* \mathfrak{q}_x}{W^2}\right) + \frac{\partial}{\partial y}\left(\frac{\mu^* \mathfrak{q}_y}{W^2}\right) + \frac{\partial}{\partial z}\left(\frac{\mu^* \mathfrak{q}_z}{W^2}\right) + \frac{\partial}{\partial t}\left(\frac{\mu^*}{W^2}\right)\right\}$$~~

– – – – – – – – – – – – – – – – – – – –

– – – – – – – – – – – – – – – – – – – –

~~$$-\frac{i}{c}\eta = -\frac{i}{c}\frac{\partial p}{\partial t}$$~~

~~$$-f_x = \frac{\partial p}{\partial x} + \frac{\mu^*}{1-\frac{\mathfrak{q}^2}{c^2}}\left\{\mathfrak{q}_x \frac{\partial \mathfrak{q}_x}{\partial x} + \mathfrak{q}_y \frac{\partial \mathfrak{q}_x}{\partial y} + \mathfrak{q}_z \frac{\partial \mathfrak{q}_x}{\partial z} + \frac{\partial \mathfrak{q}_x}{\partial t}\right\} + \mathfrak{q}_x A$$~~

~~$$+ \mathfrak{q}_x\left\{\frac{\partial}{\partial x}\left(\frac{\mu^* \mathfrak{q}_x}{1-\frac{\mathfrak{q}^2}{c^2}}\right)\right.$$~~

– – – – – – – – – – – – – – – – – – – –

– – – – – – – – – – – – – – – – – – – –

~~$$-\frac{1}{c^2}\eta = -\frac{1}{c^2}\frac{\partial p}{\partial t} + A,$$~~

~~wobei zur Abkürzung~~

~~$$A = \frac{\partial}{\partial x}\left(\frac{\mu^* \mathfrak{q}_x}{1-\frac{\mathfrak{q}^2}{c^2}}\right) + \frac{\partial}{\partial y}\left(\frac{\mu^* \mathfrak{q}_y}{1-\frac{\mathfrak{q}^2}{c^2}}\right) + \frac{\partial}{\partial z}\left(\frac{\mu^* \mathfrak{q}_z}{1-\frac{\mathfrak{q}^2}{c^2}}\right) + \frac{\partial}{\partial t}\left(\frac{\mu^*}{1-\frac{\mathfrak{q}^2}{c^2}}\right)$$~~

~~$$\mu^* = \mu + \frac{p}{c^2}$$~~

~~gesetzt ist.~~

$$\left.\begin{array}{l} -K_1 = \dfrac{\partial p}{\partial x} + \dfrac{\partial}{\partial x}(\lambda \mathfrak{q}_x \mathfrak{q}_x) + \dfrac{\partial}{\partial y}(\lambda \mathfrak{q}_x \mathfrak{q}_y) + \dfrac{\partial}{\partial z}(\lambda \mathfrak{q}_x \mathfrak{q}_z) + \dfrac{\partial}{\partial t}(\lambda \mathfrak{q}_x) \\ - \\ - \\ -i\dfrac{K_4}{c} = \dfrac{1}{c^2}\dfrac{\partial p}{\partial t} - \dfrac{\partial}{\partial x}(\lambda \mathfrak{q}_x) - \dfrac{\partial}{\partial y}(\lambda \mathfrak{q}_y) - \dfrac{\partial}{\partial z}(\lambda \mathfrak{q}_z) - \dfrac{\partial \lambda}{\partial t}. \end{array}\right\} \quad (85a)$$

Here we have set

$$\lambda = \frac{\mu + \dfrac{p}{c^2}}{1 - \dfrac{q^2}{c^2}}.$$

The quantity λ plays the role of the density, as referred to the coordinate system, while μ denotes the "rest-density," i.e., the density from the standpoint of the comoving observer.

K_1 denotes the momentum imparted by the liquid per unit volume and unit time in the direction of the X-axis, and thus $-K_1$ denotes the force exerted on the liquid from the outside per unit volume. It follows further from (85) and the meaning of $(T_{\mu\nu})$ that icK_4 is equal to the work transferred to the liquid by the external forces per unit volume and unit time. Hence we have, if we denote the volume force exerted on the liquid from the outside by $\mathfrak{f}$,

$$\mathfrak{f}_x = -K_1$$

$$-\frac{(\mathfrak{f}\mathfrak{q})}{c^2} = -\frac{iK_4}{c}$$

If we rewrite the first three equations of (85a) with the help of the fourth, we obtain the equations

$$\left.\begin{array}{r} \mathfrak{f} - \dfrac{1}{c^2}\mathfrak{q}\,(\mathfrak{f}\mathfrak{q}) = \operatorname{grad} p + \lambda\dfrac{d\mathfrak{q}}{dt} + \dfrac{1}{c^2}\mathfrak{q}\dfrac{\partial p}{\partial t} \\ \dfrac{(\mathfrak{f}\mathfrak{q})}{c^2} = \dfrac{\partial \lambda}{\partial t} + \operatorname{div}(\lambda \mathfrak{q}) - \dfrac{1}{c^2}\dfrac{\partial p}{\partial t}, \end{array}\right\} \quad (85b)$$

where we have set

$$\frac{d}{dt} = \frac{\partial}{\partial t} + \mathfrak{q}_x\frac{\partial}{\partial x} + \mathfrak{q}_y\frac{\partial}{\partial y} + \mathfrak{q}_z\frac{\partial}{\partial z}$$

in the usual manner. The first of equations (85b) corresponds to Euler's equations, and the last one to the continuity equation, which expresses at the same time the law of the conservation of energy.

$$
\left.
\begin{aligned}
-K_1 &= \frac{\partial p}{\partial x} + \frac{\partial}{\partial x}(\lambda q_x q_x) + \frac{\partial}{\partial y}(\lambda q_x q_y) + \frac{\partial}{\partial z}(\lambda q_x q_z) + \frac{\partial}{\partial t}(\lambda q_x) \\
&- - - - - - - - - - - - - - - - - - \\
&- - - - - - - - - - - - - - - - - - \\
\frac{iK_4}{c} &= \frac{1}{c^2}\frac{\partial p}{\partial t} - \frac{\partial}{\partial x}(\lambda q_x) - \frac{\partial}{\partial y}(\lambda q_y) - \frac{\partial}{\partial z}(\lambda q_z) - \frac{\partial \lambda}{\partial t}.
\end{aligned}
\right\} \quad (85a)
$$

Dabei ist

$$\lambda = \frac{\mu + \frac{p}{c^2}}{1 - \frac{q^2}{c^2}}$$

gesetzt. Die Grösse λ ~~ist~~ spielt die Rolle der aufs Koordinatensystem bezogenen Dichte, während μ die „Ruhedichte", d. h. die Dichte vom Standpunkte des mitbewegten Beobachters bedeutet.

K_1 bedeutet den im Sinne der X Achse von der Flüssigkeit pro Volumen und Zeiteinheit abgegebenen Impuls, $-K_1$ also die pro Volumeneinheit ~~Dichte der~~ auf die Flüssigkeit von aussen wirkenden Kraft. Es folgt ferner aus (85) und der Bedeutung von $(T_{\mu\nu})$, dass icK_4 gleich der pro Volumen und Zeiteinheit von den äusseren Kräften auf die Flüssigkeit übertragenen Arbeit ist. Also ist, wenn wir ~~die~~ die auf die Flüssigkeit von aussen wirkende Volumkraft mit f bezeichnen

$$f_x = -K_1$$

~~$(fq) = icK_4$~~

$$-\frac{(fq)}{c^2} = -i\frac{K_4}{c}$$

Formt man mit Hilfe der vierten der Gleichungen (85a) die drei ersten um, so erhält man die Gleichungen:

~~$f_x - \frac{1}{c^2}$~~

$$
\left.
\begin{aligned}
f - \frac{1}{c^2} q (fq) &= \text{grad}\, p + \lambda \frac{dq}{dt} + \frac{1}{c^2} q \frac{\partial p}{\partial t} \\
\frac{(fq)}{c^2} &= \frac{\partial \lambda}{\partial t} + \text{div}(\lambda q) - \frac{1}{c^2}\frac{\partial p}{\partial t}
\end{aligned}
\right\} \quad (85b)
$$

wobei in üblicher Weise

$$\frac{d}{dt} = \frac{\partial}{\partial t} + q_x \frac{\partial}{\partial x} + q_y \frac{\partial}{\partial y} + q_z \frac{\partial}{\partial z}$$

gesetzt ist. Die erste der Gleichungen (85b) entspricht den Euler'schen Gleichungen, die letzte der Kontinuitätsgleichung, welche gleichzeitig den Satz von der Erhaltung der Energie ausspricht.

BIBLIOGRAPHY

Abraham 1909 — Abraham, Max. "Zur Elektrodynamik bewegter Körper." *Circolo Matematico di Palermo. Rendiconti* 28 (1909): 1–28.

Abraham/Föppl 1904 — Abraham, Max. *Theorie der Elektrizität.* Vol. 1, August Föppl, *Einführung in die Maxwellsche Theorie der Elektrizität.* 2d rev. ed. Max Abraham, ed. Leipzig: Teubner, 1904.

Blumenthal 1920 — Blumenthal, Otto. *Das Relativitätsprinzip: eine Sammlung von Abhandlungen.* 3rd rev. ed. Leipzig: Teubner, 1920.

De Sitter 1913 — De Sitter, Willem. "Ein astronomischer Beweis für die Konstanz der Lichtgeschwindigkeit." *Physikalische Zeitschrift* 14 (1913): 429.

Ehrenfest 1912 — Ehrenfest, Paul. "Zur Frage nach der Entbehrlichkeit des Weltäthers." *Physikalische Zeitschrift* 13 (1912): 317–319.

Eichenwald 1903 — Eichenwald, A. "Über die magnetischen Wirkungen bewegter Körper im elektrostatischen Felde." *Annalen der Physik* 11 (1903): 1–30, 421–441.

Eichenwald 1904 — ———. "Über die magnetischen Wirkungen bewegter Körper im elektrostatischen Felde (Nachtrag)." *Annalen der Physik* 13 (1904): 919–943.

Einstein 1905 — Einstein, Albert. "Ist die Trägheit eines Körpers von seinem Energieinhalt abhängig?" *Annalen der Physik* 18 (1905): 639–641.

Einstein 1907a — ———. "Über die Möglichkeit einer neuen Prüfung des Relativitätsprinzips." *Annalen der Physik* 23 (1907): 197–198

Einstein 1907b — ———. "Über das Relativitätsprinzip und die aus demselben gezogenen Folgerungen." *Jahrbuch der Radioaktivität und Elektronik* 4 (1907): 411–462.

Einstein 1910 — ———. "Le principe de relativité et ses conséquences dans la physique moderne." *Archives des sciences physiques et naturelles* 29 (1910): 5–28, 125–144.

Einstein 1911 — ———. "Die Relativitäts-Theorie." *Naturforschende Gesellschaft in Zürich. Vierteljahrsschrift* 56 (1911): 1–14.

Einstein 1912a — ———. "Lichtgeschwindigkeit und Statik des Gravitationsfeldes." *Annalen der Physik* 38 (1912): 355–369.

Einstein 1912b — ———. "Relativität und Gravitation. Erwiderung auf eine Bemerkung von M. Abraham." *Annalen der Physik* 38 (1912): 1059–1064.

Einstein 1921 — ———. "Geometrie und Erfahrung." *Königlich Preußsche Akademie der Wissenschaften* (Berlin). *Sitzungsberichte* (1921): 123–130.

Einstein 1987 — ———. *The Collected Papers of Albert Einstein.* Vol. 1, *The Early Years: Writings, 1879–1902.* Martin Klein et al., eds. Princeton, N.J.: Princeton University Press, 1987.

Einstein 1989 — ———. *The Collected Papers of Albert Einstein,* Vol. 2, *The Swiss Years: Writings, 1900–1909.* John Stachel et al., eds. Princeton, N.J.: Princeton University Press, 1989.

Einstein 1993a — ———. *The Collected Papers of Albert Einstein,* Vol. 3, *The Swiss Years: Writings, 1909–1911.* Martin Klein et al., eds. Princeton, N.J.: Princeton University Press, 1993.

Einstein 1993b — ———. *The Collected Papers of Albert Einstein,* Vol. 5, *The Swiss Years: Correspondence, 1902–1914.* Martin Klein et al., eds. Princeton, N.J.: Princeton University Press, 1993.

Einstein 1995 — ———. *The Collected Papers of Albert Einstein,* Vol. 4, *The Swiss Years: Writings, 1912–1914.* Martin Klein et al., eds. Princeton, N.J.: Princeton University Press, 1995.

Einstein and Fokker 1914 Einstein, Albert, and Fokker, Adriaan D. "Die Nordströmsche Gravitationstheorie vom Standpunkt des absoluten Differentialkalküs." *Annalen der Physik* 44 (1914): 321–328.

Einstein and Grossmann 1913 Einstein, Albert, and Grossmann, Marcel. *Entwurf einer verallgemeinerten Relativitätstheorie und einer Theorie der Gravitation.* Leipzig: Teubner, 1913. Reprinted as *Einstein and Grossmann 1914a.*

Einstein and Grossman 1914a ———."Entwurf einer verallgemeinerten Relativitätstheorie und einer Theorie der Gravitation." *Zeitschrift für Mathematik und Physik* 62 (1914): 225–259.

Einstein and Grossmann 1914b ———. "Kovarianzeigenschaften der Feldgleichungen der auf die verallgemeinerte Relativitätstheorie gegründeten Gravitationstheorie." *Zeitschrift für Mathematik und Physik* 63 (1914): 215–225.

Einstein and Laub 1908a Einstein, Albert, and Laub, Jakob. "Über die elektromagnetischen Grundgleichungen für bewegte Körper." *Annalen der Physik* 26 (1908): 532–540.

Einstein and Laub 1908b ———. "Über die im elektromagnetischen Felde auf ruhende Körper ausgeübten ponderomotorischen Kräfte." *Annalen der Physik* 26 (1908): 541–550.

Fizeau 1851 Fizeau, Armand H. "Sur les hypothèses relatives à l'éther lumineux, et sur une expérience qui paraît démontrer que le mouvement des corps change la vitesse avec laquelle la lumière se propage dans leur intérieur." *Académie des sciences* (Paris). *Comptes rendus* 33 (1851): 349–355.

Fölsing 1993 Fölsing, Albrecht. *Albert Einstein: Eine Biographie.* Frankfurt/M: Suhrkamp Verlag, 1993.

Föppl 1894 Föppl, August. *Einführung in die Maxwell'sche Theorie der Elektricität.* Leipzig: Teubner, 1894. (For 2d ed., see *Abraham/Föppl 1904.)*

French 1979 French, A. P., ed. *Einstein: A Centenary Volume.* Cambridge, Massachusetts: Harvard University Press, 1979.

Heaviside 1892 Heaviside, Oliver. *Electrical Papers.* Vol. 1. London: Macmillan, 1892.

Ives and Stilwell 1938 Ives, Herbert E., and Stilwell, G. R. "An Experimental Study of the Rate of a Moving Atomic Clock." *Journal of the Optical Society of America* 28 (1938): 215–226.

Kowalewski 1909 Kowalewski, Gerhard. *Einführung in die Determinantentheorie einschließlich der unendlichen und der Fredholmschen Determinanten.* Leipzig: Veit, 1909.

Laue 1907 Laue, Max. "Die Mitführung des Lichtes durch bewegte Körper nach dem Relativitätsprinzip." *Annalen der Physik* 23 (1907): 989–990.

Laue 1911 ———. *Das Relativitätsprinzip.* Braunschweig: Vieweg, 1911.

Laue 1913 ———. *Das Relativitätsprinzip.* 2d rev. ed. Braunschweig: Vieweg, 1913.

Lorentz 1892 Lorentz, Hendrik A. "De relatieve beweging van de aarde en den aether." *Koninklijke Akademie van Wetenschappen* (Amsterdam). *Wis- en Natuurkundige Afdeeling. Verslagen der Zittingen 1* (1892–1893): 74–79.

Lorentz 1904a ———. "Maxwells elektromagnetische Theorie." In *Encyklopädie der mathematischen Wissenschaften, mit Einschluss ihrer Anwendungen.* Vol. 5, *Physik,* part 2, pp. 63–144. Arnold Sommerfeld, ed. Leipzig: Teubner, 1904–1922. Issued 16 June 1904.

Lorentz 1904b ———. "Weiterbildung der Maxwellschen Theorie. Elektronentheorie." In *Encyklopädie der mathematischen Wissenschaften, mit Einschluss ihrer Anwendungen.* Vol. 5, *Physik,* part 2, pp. 145–280. Arnold Sommerfeld, ed. Leipzig: Teubner, 1904–1922. Issued 16 June 1904.

Lorentz 1909 ———. *The Theory of Electrons and lts Applications to the Phenomena of Light and Radiant Heat.* Leipzig: Teubner, 1909.

McCormmach 1973 McCormmach, Russell. "Lorentz, Hendrik Antoon." In *Dictionary of Scientific Biography.* Charles C. Gillispie, ed. Vol. 8, pp. 487–500. New York: Scribner's, 1973.

Mach 1908 Mach, Ernst. *Die Mechanik in ihrer Entwickelung. Historisch-kritisch dargestellt.* 6th rev. ed. Leipzig: Brockhaus, 1908.

Maxwell 1891 Maxwell, James Clerk. *A Treatise on Electricity and Magnetism.* 2 vols. 2d ed. Oxford: Clarendon Press, 1891.

Miller 1981 Miller, Arthur I. *Albert Einstein's Special Theory of Relativity: Emergence (1905) and Early Interpretation (1905–1911).* Reading, Mass.: Addison-Wesley, 1981.

Minkowski 1908 Minkowski, Helmann. "Die Grundgleichungen für die elektromagnetischen Vorgänge in bewegten Körpern." *Königliche Gesellschaft der Wissenschaften zu Göttingen. Mathematisch-physikalische Klasse. Nachrichten* (1908): 53–111. Reprinted in *Minkowski 1911,* vol. 2, pp. 352–404.

Minkowski 1909 ———. *Raum und Zeit. Vortrag gehalten auf der 80. Naturforscher-Versammlung zu Köln am 21. September 1908.* Leipzig: Teubner, 1909. Also printed in *Physikalische Zeitschrift* 10 (1909): 104–111, and reprinted in *Minkowski 1911,* vol. 2, pp. 431–444.

Minkowski 1911 ———. *Gesammelte Abhandlungen.* David Hilbert, ed. 2 vols. Leipzig: Teubner, 1911. Reprint, New York: Chelsea, 1967.

Minkowski/Born 1910 Minkowski, Hermann. "Eine Ableitung der Grundgleichungen für die elektromagnetischen Vorgänge in bewegten Körpern vom Standpunkte der Elektronentheorie" [prepared for publication by Max Born]. *Mathematische Annalen* 68 (1910): 526–551. Reprinted in *Minkowski 1911,* vol. 2, pp. 405–430.

Pais 1982 Pais, Abraham. *'Subtle is the Lord . . .': The Science and the Life of Albert Einstein.* Oxford & New York: Clarendon Press, 1982.

Planck 1907 Planck, Max. "Zur Dynamik bewegter Systeme." *Königlich Preußische Akademie der Wissenschaften* (Berlin). *Sitzungsberichte* (1907): 542–570. Reprinted in *Annalen der Physik* 26 (1908): 1–34; and in *Planck 1958,* vol. 2, pp. 176–209.

Planck 1958 ———. *Physikalische Abhandlungen und Vorträge.* 3 vols. Braunschweig: Wieweg, 1958.

Ritz 1908a Ritz, Walter. "Recherches critiques sur l'électrodynamique générale." *Annales de chimie et de physique* 13 (1908): 145–275.

Ritz 1908b ———. "Über die Grundlagen der Elektrodynamik und die Theorie der schwarzen Strahlung." *Physikalische Zeitschrift* 9 (1908): 903–907.

Röntgen 1888 Röntgen, Wilhelm Conrad. "Ueber die durch Bewegung eines im homogenen electrischen Felde befindlichen Dielectricums hervorgerufene electrodynamische Kraft." *Annalen der Physik und Chemie* 35 (1888): 264–270.

Rowland 1878 Rowland, Henry A. "On the Magnetic Effect of Electric Convection." *The American Journal of Science and Arts* 15 (1878): 30–38.

Rowland and Hutchinson 1889 Rowland, Henry A., and Hutchinson, Cary T. "On the Electromagnetic Effect of Convection-Currents." *Philosophical Magazine* 27 (1889): 445–460.

Shankland 1963 Shankland, R. S. "Conversations with Albert Einstein." *American Journal of Physics* 31 (1963): 47–57.

Siegel 1991 Siegel, Daniel M. *Innovation in Maxwell's Electromagnetic Theory: Molecular Vortices, Displacement Current, and Light.* Cambridge: Cambridge University Press, 1991.

Sommerfeld 1910a Sommerfeld, Arnold. "Zur Relativitätstheorie I. Vierdimensionale Vektoralgebra." *Annalen der Physik* 32 (1910): 749–776.

Sommerfeld 1910b ———. "Zur Relativitätstheorie II. Vierdimensionale Vektoranalysis." *Annalen der Physik* 33 (1910): 649–689.

Sommerfeld 1923 ———. *The Principle of Relativity: A Collection of Original Memoirs.* Arnold Sommerfeld, ed. London: Methuen, 1923.

Weil 1960 Weil, E. *Albert Einstein: A Bibliography of His Scientific Papers 1901–1954.* London, 1960.

Wilson 1904 Wilson, Harold Albert. "On the Electric Effect of Rotating a Dielectric in a Magnetic Field." *Royal Society of London. Philosophical Transactions A* 204 (1904–1905): 121–137.

Albert Einstein Archives, Hebrew University, Jerusalem (photocopies available at Princeton University Library). The dossier of correspondence with Erich Marx and the Akademische Verlagsgesellschaft, Leipzig, is in file 41.982 and after.

NOTES

[1] See *Einstein 1995*, pp. 3–6, for the arguments on which the dating of this document is based.

[2] Einstein's definition of electrical charge is similar to the definition of charge in his lecture notes for a course on electricity and magnetism at the University of Zurich. See *Einstein 1993a*, doc. 11, which in turn follows Mach's definition of mass in *Mach 1908*, sec. 5.

[3] See *Heaviside 1892*, p. 199.

[4] Einstein's use of lower-case Gothic characters for electromagnetic vectorial quantities follows *Lorentz 1904b*.

[5] See *Siegel 1991* for a historical discussion of Maxwell's introduction of the displacement current.

[6] See *Rowland 1878* and *Rowland and Hutchinson 1889*. The experiment was proposed by Maxwell (see *Maxwell 1891*, vol. 2, p. 415). See, e.g., *Laue 1911*, pp. 9–10, for a discussion.

[7] See *Maxwell 1891*, vol. 2, p. 415.

[8] In the following equation $\mathfrak{E}$ should be $\mathfrak{e}$.

[9] This deleted passage as well as other similar ones suggest that Einstein originally intended to add a section on vector calculus to his text.

[10] Einstein uses the notations $\mathfrak{a}\mathfrak{b}$, $(\mathfrak{a},\mathfrak{b})$, or $(\mathfrak{a}\ \mathfrak{b})$ for the scalar product of two vectors $\mathfrak{a}$ and $\mathfrak{b}$. The vector product is denoted by $[\mathfrak{a},\mathfrak{b}]$ or $[\mathfrak{a}\ \mathfrak{b}]$.

[11] The p_{xx} etc. in the equations below are Laue's notation (see *Laue 1911*, p. 82)

[12] See, e.g., *Lorentz 1904b*, p. 156.

[13] See *Lorentz 1904b*, sec. IV.

[14] For Lorentz's analysis of the normal Zeeman effect, see, e.g., *Lorentz 1909*, chap. 3

[15] See *Einstein 1907b*, §10 (doc. 47 in *Einstein 1989*), for Einstein's earlier discussion of experiments on cathode rays as empirical support for the theory of relativity. For a historical account, see *Miller 1981*, chap. 1.

[16] See, e.g., *McCormmach 1973* for a historical discussion of Lorentz's electron theory.

[17] See *Lorentz 1904b*, §§26-28, for a discussion.

[18] See *Röntgen 1888* and *Eichenwald 1903, 1904*. See also *Lorentz 1904a*, §17, and *Lorentz 1904b*, §34, for discussions.

[19] See *Wilson 1904*

[20] See *Lorentz 1904b*, §48.

[21] The first "grad" in the following equation should be "div."

[22] The sign of $\mathfrak{e}_x\rho$ in the following equation should be positive.

[23] The sign of $\mathfrak{e}_x\rho$ in the following equation should be negative.

[24] A subscript x is missing on the right-hand side of the following equation.

[25] For discussions of the duality of electric and magnetic phenomena, see *Föppl 1894*, p. 131, and *Lorentz 1904a*, §18.

[26] From the deleted passage at the bottom of page 13 it appears that Einstein originally planned to include a derivation of Stokes's theorem in a mathematical appendix; instead at this point in the original text he indicates a note he has appended at the foot of the page. An additional page in the original, on which he continues the note under the heading "(Portsetzung zur Anmerkung)," is paginated by Einstein as "13a." We have reproduced and translated this note on page 66 of this volume.

[27] See *Minkowski 1908*, §8.

[28] For the introduction of this term, see *Heaviside 1892*, pp. 446 and 546.

[29] See *Fizeau 1851*. Einstein discussed Fizeau's experiment on various earlier occasions (see, e.g., *Einstein 1911*, pp. 3-6 [doc. 17 in *Einstein 1993a*]). For a later recollection on the significance of this experiment for his formulation of the special theory of relativity, see *Shankland 1963*.

[30] Einstein probably intended to include the rule curl curl $A = \nabla^2 A - \text{grad}\,\text{div} A$ in a mathematical appendix to his text.

[31] See *Laue 1911*, pp. 33-34. Einstein and Grossmann later introduced similar terminology in the context of their generalized theory of gravitation (see *Einstein and Grossmann 1914b*, p. 221).

[32] Einstein's manuscript does not include a discussion of the Michelson-Morley experiment.

[33] For a similar discussion of Fizeau's experiment, see *Einstein 1911*, p. 4 (doc. 17 in *Einstein 1993a*).

[34] A similar formulation is found in *Einstein 1912b*, p. 1059 (doc. 8 in *Einstein 1995*). This paper was received by the editor of the *Annalen der Physik* on 4 July 1912.

[35] Einstein's consideration of a possible role played by the Earth together with all other objects is similar to his introduction of the system U, representing "the totality of all other systems in the world" ("die Gesamtheit aller übrigen Systeme der Welt"), in *Einstein 1912b*, p. 1060 (doc. 8 in *Einstein 1995*).

[36] Einstein considered gravitational effects on the speed of light in his contemporary studies on gravitation (see *Einstein 1995*, the editorial note, "Einstein on Gravitation and Relativity: The Static Field," sec. II).

[37] Einstein inserted the text on [p.20a] to replace the deleted passages at the foot of [p. 20] and at the top of [p.21].

[38] "Im Falle I" should be "Im Falle II."

[39] As becomes clear from the discussion in the remainder of the paragraph, Einstein refers to the Dutch astronomer Willem de Sitter, who had found that Ritz's hypothesis is incompatible with existing observational data on double stars (see *De Sitter 1913*,

which was published on 1 May 1913). See also Einstein's comments on De Sitter's paper in Einstein to Paul Ehrenfest, 28 May 1913 (*Einstein 1993b*, doc. 441) and the discussion of this passage in the editorial note, "Einstein's Manuscript on the Special Theory of Relativity," sec. III.

[40] "Im Falle II" should be "Im Falle I."

[41] See *Laue 1907*.

[42] See *Ritz 1908a, 1908b* and *Ehrenfest 1912*. This passage is discussed im the editorial note, "Einstein's Manuscript on the Special Theory of Relativity," sec. III in *Einstein 1995*; see also the editorial note, "Einstein on the Theory of Relativity," p. 263, in *Einstein 1989*, for more on Ritz's theory.

[43] These complications are discussed in Einstein's correspondence with Paul Ehrenfest in spring and early summer 1912 (see *Einstein 1993b* docs. 384, 390, 394, 404, and 409).

[44] Einstein inserted the text on [p. 20a] to replace the deleted passages ending here. This page is written in darker ink, suggesting that it was added at a later time.

[45] The reference to rotation is probably an allusion to the problem of a rotating disk discussed in *Einstein 1912a*, p. 356 (doc. 3 in *Einstein 1995*).

[46] Einstein later referred to this approach as "practical geometry" ("praktische Geometrie"; see *Einstein 1921*, p. 125).

[47] For a similar discussion of the concept of a clock, see, e.g., *Einstein 1910*, p. 21 (doc 2 in *Einstein 1993a*).

[48] "$t_A - t_0 = t_0' - t_A$" should be $t_A - t_0 = t_0' - t_A'$. In *Laue 1911* Laue does not consider two signals propagating in opposite directions. In the second edition of his book, however, he does stress independence of the direction of the signal (see *Laue 1913*, p. 37).

[49] *Minkowski 1908* uses $x_4 = ict$; *Laue 1911* introduces $u = ct$, $l = ict$.

[50] A similar representation is used in *Laue 1911*.

[51] Initially, Einstein was less enthusiastic about the four dimensional formulation of special relativity (see *Einstein and Laub 1908a*, in which its use was avoided), but in summer 1910 his opinion had changed (see Einstein to Arnold Sommerfeld, July 1910 [doc. 211 in *Einstein 1993b*]).

[52] See *Lorentz 1892*.

[53] This test was first proposed in *Einstein 1907a* (doc. 41 in *Einstein 1989*) but successfully performed only much later (see *Ives and Stilwell 1938*).

[54] See *Laue 1907*.

[55] The equations below are the last in the manuscript in which Einstein uses the notation "curl." In the remainder the operation is denoted by "rot."

[56] See §9.

[57] The argument of this paragraph follows *Einstein 1905* (doc. 24 in *Einstein 1989*).

[58] See *Einstein 1907b*, pp. 442–43 (doc. 47 in *Einstein 1989*), for similar calculations.

[59] In a letter to Wilhelm Wien of 10 July 1912 (doc. 413 in *Einstein 1993b*) Einstein inquired whether a test of the proportionality of inertial and gravitational mass for radioactive bodies was experimentally feasible. The importance of such a test is also emphasized in *Einstein 1912b*, p. 1062 (doc. 8 in *Einstein 1995*).

[60] See the historical discussion of contemporary experiments in *Einstein 1989*, the editorial note, "Einstein on the Theory of Relativity," pp. 270–271.

[61] Beginning with the words "Vektor-und Tensoren-Theorie," the remainder of this heading is written in dark ink and replaces the deleted word "Geometrie." For a discussion of the relationship between geometry and tensor calculus as tools for Einstein's development of a generalized theory of relativity, see the editorial note, "Einstein on Gravitation and Relativity: The Collaboration with Marcel Grossmann" in *Einstein 1995*.

[62] "$\alpha_{31}y$" in the third equation below should be "$\alpha_{32}y$."

[63] See, e.g., the discussion in *Minkowski 1909*.

[64] These questions are not literal quotations from Minkowski's writings.

[65] For Minkowski's characterization of the laws of physics in a four-dimensional world, see *Minkowski 1908*, p. 57, where he states: "The entire world appears to be resolved into such world lines, and I would like to state at the outset that in my view the physical laws should find their most perfect expression in the interrelations among these world lines" ("Die ganze Weldt erscheint aufgelöst in solche Weltlinien, und ich möchte sogleich vorwegnehmen, dass meiner Meinung nach die physikalischen Gesetze ihren vollkommensten Ausdruck als Wech selbeziehungen unter diesen Weltlinien finden dürften").

[66] See *Einstein and Grossmann 1913*, part 2 (doc. 13 in *Einstein 1995*), for a systematic treatment of tensor calculus. For a brief discussion of the contemporary understanding of tensors as well as of Einstein's and Grossmann's contribution to tensor calculus, see the editorial note, "Einstein's Research Notes on a Generalized Theory of Relativity," sec 6. II and III, in *Einstein 1995*.

[67] Einstein follows Sommerfeld's rather than Minkowski's terminology (see *Sommerfeld 1910a*, p. 750). This terminology is also used in *Laue 1911*.

[68] For Sommerfeld's definition of an axial vector, see *Sommerfeld 1910a*, p. 750; for Laue's definition, see *Laue 1911*, p. 60.

[69] See *Laue 1911*, p 61.

[70] From this point on the manuscript is written on paper of Swiss manufacture.

[71] Einstein's use of parentheses to distinguish between vectors (or tensors) and their components is not common in the contemporary literature, including Einstein's own writings. For exceptions, see *Einstein and Fokker 1914* , pp. 322–323 (doc. 28 in *Einstein 1995*), and Einstein's lecture notes for his course on electricity and magnetism at the ETH in winter semester 1913/1914 (doc. 19 in *Einstein 1995*).

[72] Einstein's general definition of a tensor implies the existence of sixteen rather than of ten independent components. It was, however, customary at the time to restrict the notion of a tensor to what are now called symmetric, second-rank tensors (see *Sommerfeld 1910a*, p. 767).

[73] Neither Minkowski, nor Sommerfeld, nor Laue makes use of a general tensor concept as it is defined here.

[74] This notation is neither common in the literature of the time nor is it found in other contemporary writings by Einstein.

[75] For a systematic discussion of vectors as second-rank, antisymmetric tensors, see *Einstein and Grossmann*, part 2, §3 (doc. 13 in *Einstein 1995*).

[76] "ein Tensor" should be "eines Tensors."

[77] See *Sommerfeld 1910a*, p. 753.

[78] This notation is neither common in the literature of the time nor is it found in other contemporary writings by Einstein.

[79] For contemporary expositions of the theory of determinants, see, e.g., *Minkowski 1908* and *Kowalewski 1909.*

[80] See *Laue 1911a,* p. 74. In *Einstein and Grossmann 1913*, p. 21 (doc. 13 in Einstein 1995), this scalar is called "Laue's scalar" ("Lauescher Skalar"); Einstein also states that Laue has directed his attention to its existence.

[81] "T_{ik}" should be "$(\dot{T}^*_{ik})$."

[82] Einstein's general definition of this operation is not common in the contemporary physics literature.

[83] This generalization of the Laplace operator is also found in *Sommerfeld 1910b*, p. 661, and *Laue 1911*, p. 71.

[84] This section, which contains Einstein's first treatment of four-dimensional electrodynamics, is largely based on earlier work by Minkowski, Sommerfeld, and Laue (see *Minkowski 1908, Sommerfeld 1910a, 1910b,* and *Laue 1911).* For an earlier comment by Einstein on Sommerfeld's contribution, see Einstein to Arnold Sommerfeld, July 1910 (doc. 211 in *Einstein 1993b*).

[85] In 1908 Einstein and Jakob Laub had published two papers on the electrodynamics of moving media (see *Einstein and Laub 1908a, 1908b* [docs. 51 and 52 in *Einstein 1989*]). See *Einstein 1989,* the editorial note, "Einstein and Laub on the Electrodynamics of Moving Media," pp. 503-507, for a historical discussion.

[86] Lorentz discusses the electrodynamics of ponderable bodies in *Lorentz 1904b,* §§26-55; for Minkowski's work, see *Minkowski 1908, 1909,* and *Minkowski/Born 1910;* for Abraham's contribution, see *Abraham 1909;* Laue includes a treatment of this subject in *Laue 1911* (see in particular pp. 127–129). The electrodynamics of moving media is frequently mentioned in Einstein's contemporary correspondence, in particular with Jakob Laub (see *Einstein 1993b*). For Abraham's comparison of the various approaches to this problem according to their being more or less "intuitive" ("anschaulich"), see *Abraham 1909,* p. 3.

[87] See *Minkowski 1908, p. 91.* See also the discussion of the relationship between Minkowski's contribution and Einstein's earlier work on this subject in *Einstein 1989,* the editorial note, "Einstein and Laub on the Electrodynamics of Moving Media," pp. 503–507.

[88] For the definition of the fundamental quantities occurring in Minkowski's equations, see *Minkowski 1908,* §7.

[89] See *Planck 1907, §2; Minkowski 1908,* pp. 93, 97–98; *Abraham 1909,* §3; and *Laue 1911,* p. 160.

[90] For the expression "thread" ("Faden"), see *Minkowski 1908,* p. 100.

[91] Einstein refers to the last formula of §2.

[92] Einstein's discussion of relativistic hydrodynamics in his course on electricity and magnetism at the ETH in winter semester 1913/1914 (see *Einstein 1995,* doc. 19 for Einstein's notes) parallels the treatment presented here. For a contemporary review of this subject, see *Laue 1913,* §§36, 37.

[93] "K_μ" in the equation below should be "K_ν".

[94] The notation $w = \sqrt{1-\beta^2}$ is also used in Einstein's course on electricity and magnetism at the ETH in winter semester 1913/1914 (see *Einstein 1995,* doc. 19).